11th Process Plant Safety Symposium 2009

Topical Conference at the 2009 AIChE Spring National Meeting

Tampa, Florida
26 – 30 April 2009

ISBN: 978-1-61567-254-7

Printed from e-media with permission by:

Curran Associates, Inc.
57 Morehouse Lane
Red Hook, NY 12571

Some format issues inherent in the e-media version may also appear in this print version.

Printed by Curran Associates, Inc. (2009)

AIChE
3 Park Avenue
New York, NY 10016-5991

Phone: (203) 702-7660
Fax: (203) 775-5177

www.aiche.org

Additional copies of this publication are available from:

Curran Associates, Inc.
57 Morehouse Lane
Red Hook, NY 12571 USA
Phone: 845-758-0400
Fax: 845-758-2634
Email: curran@proceedings.com
Web: www.proceedings.com

TABLE OF CONTENTS

PROCESS PLANT SAFETY SYMPOSIUM

Welcome to the 11th Process Plant Safety Symposium. The first Process Plant Safety Symposium (PPSS) was held in 1992. The PPSS was originally organized by the South Texas section of AIChE. Today, it continues to focus on process safety and risk challenges and solutions for immediate application in process plants throughout our industry. PPSS is now an annual symposium and is part of the AIChE Global Congress on Process Safety. PPSS sessions and focus areas this year include:

- **Global Congress Keynote Session** (Joint Session with LPS and CCPS)
- **Applications of LOPA (Layer of Protection Analysis):** Layer of Protection Analysis (LOPA) continues to be a powerful and popular tool for assessing risk. This session contains papers focusing on applied principles of Layer of Protection Analysis. Special consideration was given to papers that focus on actual plant experiences with LOPA implementations, both positive and negative.
- **Safety Culture and Operational Discipline:** This session contains papers covering the cultural aspects of a successful Process Safety program. The papers present new and innovative ideas on corporate-wide and site-specific safety culture improvements, on the development or improvement in operational discipline, and on the "human element" in operating safely.
- **Mechanical Integrity - State of the Art:** This session contains papers on the latest mechanical integrity concepts, practices and programs. Topics include equipment inspection, interlock testing, reliability initiatives, and overall system design. An additional area of focus is on the challenges of meeting the needs of a global organization, including language & cultural barriers, widely varied national regulations and availability of qualified resources
- **Process Safety – from the Top-Down and Bottom-up:** The theme of the papers in this session is implementing Process Safety systems from two different perspectives - Senior Management and Technical Personnel. The perspective of the former is what they need from process safety and risk practitioners to effectively facilitate decision making. The perspective of the latter is what they need to gain ongoing positive support of management. The session provides case histories, methods of analysis and assessment, approaches to presenting and "selling" general or specific process safety needs.
- **Applications of Risk Analysis Techniques:** Hazard identification and risk assessment continue to be a critical part of the process safety efforts in process plants. The papers in this session highlight methods, tools and examples of successful application or lessons learned from misapplication in a plant environment.
- **Case Histories and Lessons Learned** (Joint Session with LPS and CCPS) is always a favored session reviewing accidents that have occurred and the lessons we can learn. The presentations will include cryogenic plant, flammable chemical and reactor runaway excursions.
- **Global Congress on Process Safety Poster Session**. This session will be held in conjunction with the Global Congress on Process Safety reception. Nearly 20 different poster presentations on many diverse topics will be on display. Come enjoy the reception and the posters. Take advantage of the opportunity to talk with the authors one-on-one.

There are many opportunities offered by these papers to extend your knowledge of process safety and risk applications. The intent of all of these papers is to provide immediately applicable solutions to the challenges that you face in your professional practice and process plants. Copies of these papers are generally contained on this CD for your reference.

We thank all who have contributed their time and effort to make this a successful symposium, including the session chairs and co-chairs, the organizing committee, presenters and authors. We also thank you for joining us in Tampa.

Enjoy the symposium and all it has to offer you with the latest innovations in process safety.

John Champion
PPSS Chair

Colin Howat
PPSS Co-Chair

2009 Global Congress on Process Safety

5TH GLOBAL CONGRESS ON PROCESS SAFETY

Plenary and Luncheon Speakers

2009 Global Congress Opening and Keynote Address, April 27, 8:30AM-9:30AM

- The Global Congress Keynote Address by Invited Speaker, **John Bresland**, Chairman and CEO of the U.S. Chemical Safety Board.
- *Ballroom A, Tampa Convention Center*

5th Global Congress on Process Safety Keynote Luncheon, April 27 11:45AM-1:15PM

- Luncheon Keynote will be given by **Laura Itle,** Technical Lead for the Office of SAFETY Act Implementation. Dr. Itle will give a presentation titled "SAFETY Act and You".
- *Ballroom C, Tampa Convention Center*

5th Global Congress on Process Safety Luncheon Talk, April 29, 12:00PM-1:15PM

- **Lisa Morrison**, PSM Audit Manager of BP International, will discuss BP's PSM Audit Program.
- *Ballroom C, Tampa Convention Center*

MONDAY, APRIL 27

Session #16 *Joint Plenary Session (CCPS/LPS/PPSS)*

8:00 am Introductions & Welcome, **John Murphy** and **Scott Berger**

8:10 am Introductory Remarks
- **Jean-Paul Lacoursiere**, LPS Chair
- **John Champion**, PPSS Chair
- **Eric Freiburger**, CCPS Chair

8:20 am William H. Doyle LPS Best Paper Award

8:30 am Keynote Address: **Invited Speaker, Mr. John Bresland**, Chairman and CEO, U.S. Chemical Safety & Hazard Investigation Board

9:20 am **Coffee and Networking Break in Exhibit Hall**

	43rd Annual Loss Prevention Symposium	11th Process Plant Safety Symposiun	24th CCPS International Conference
	Session #18: **Fire, Explosion and Reactive Chemicals** Location: Room 24/25 **Session Co-Chairs: Stanley S. Grossel & John A. Noronha**	***Session #19:*** **Applications of LOPA** Location: Room 22/23 **Session Co-Chairs: Jack Chosnack & James Gary**	***Session #20:*** **Auditing Process Safety** Location: Ballroom D **Session Chair: Christie Arseneau**
10:00 am	Approach for the Development of a More Efficient and Safer Process in the Pharmaceutical Industry *(L. R. Saenz)*	Key Issues with Implementing LOPA (Layer of Protection Analysis) – Perspective from One of the Originators of LOPA *(Bill Bridges)*	Can Auditing Be Improved to Drive PSM Performance Gains? *(James A. Klein)*
10:30 am	Predicting Explosibility Properties of Chemicals from Quantitative Structure — Property Relationships *(Patricia Rotureau)*	Experience of a Very Large Layer of Protection Analysis with More than 2000 'Cause – Consequence Pairs' Covering a Large Gas Terminal in Europe *(Richard Gowland)*	Auditing in Four Levels to Assure Completeness *(Adrian L. Sepeda)*
11:00 am	Minimum Ignition Energy of Dust Cloud. The Effect of the Amount of Inert Dust *(Luca Marmo)*	Improved Integration of LOPA with Hazop Analyses *(Dick Baum)*	Auditing Global Logistics Operations – A Process Safety Focus *(Michael R. Green)*
11:45 am	Luncheon With Speaker: **Dr. Laura Itle**, Officer of Safety ACT Implementation		
	Session #24: **Consequence Modeling** Location: Room 24/25 **Session Co-Chairs: David G. Clark & Brian R. Dunbobbin**	***Session #25:*** **Safety Culture and Operating Discipline** Location: Room 22/23 **Session Co-Chairs: James R. Thompson & James A. Klein**	***Session #23:*** **Implementing PSM Globally** Location: Ballroom D **Session Chair: Shakeel Kadri**
1:30 pm	Development of Algorithms for Predicting Ignition Probabilities and Explosion Frequencies *(Mike Moosemiller)*	The Impact of the BP Baker Report *(Stephanie Payne)*	Training Tools to Support a Global Process Safety Culture *(Dan Reinke)*
2:00 pm	On the Use of Models for Accident Investigations *(Kees van Wingerden)*	Identifying "Culture Issues" Using Root Cause Analysis *(Mark Paradies)*	Implementing and Sustaining a Global Process Safety Program *(Christie A. Arseneau)*
2:30 pm	A Versatile Approach to Explosion Vent Design for Enclosures Handling Flammable Vapors *(David Herrmann)*	Implementing a Culture of Safety and Operations Excellence at Newport Chemical Demilitarization Facility (NECDF) *(Tulanda Brown)*	Implementing PSM Globally *(David Cummings)*
3:00 pm	**Coffee and Networking Break in Exhibit Hall**		

MONDAY, APRIL 27 Continued

	Session #24: **Consequence Modeling (Continued)** Location: Room 24/25 **Session Co-Chairs: David G. Clark & Brian R. Dunbobbin**	***Session #25:*** **Safety Culture and Operating Discipline (Continued)** Location: Room 22/23 **Session Co-Chairs: James R. Thompson & James A. Klein**	***Session #41:*** **Tools to ID Hazards** Location: Ballroom D **Session Chair: Robert F. Wasileski**
3:30 pm	Consequence Modeling Lack of Accuracy and its Effects on Quantitative Risk Analysis Results *(Mohammed J. Ayyoubi)*	From Training to Human Reliability: Lessons Learned During the Implementation of a Human Reliability Function in the Pharmaceutical Industry *(Ginette Collazo-Ramos)*	Accident Investigation Using Process Control Event Diagrams *(D. R. Morrison)*
4:00 pm	Advanced CFD Modeling on VCE and Vapor Dispersion *(Anna Qiao)*	Improving Operating Discipline Through the Successful Implementation of a Behavior-Based Safety Program *(B. Vaughen)*	The Retro-Synthesis of Risk Assessment Tools for Hazard Evaluation and Screening Analysis *(Fred Henselwood)*
4:30 pm	LNG Modeling for Facility Siting *(David Moore)*	Safety Culture Implementation Programs in a Developing Country *(Robin Pitlado)*	A Practical Approach to Hazard Identification for Operations and Maintenance Workers *(Christy Franklyn)*

TUESDAY, APRIL 28

	43rd Annual Loss Prevention Symposium	**11th Process Plant Safety Symposiun**	**24th CCPS International Conference**
	Session #46: **Hazard Evaluation and LOPA** Location: Room 24/25 **Session Co-Chairs: Robert W. Johnson & Mike Moosemiller**	***Session #47:*** **Mechanical Integrity — State of the Art** Location: Room 22/23 **Session Co-Chairs: Steven D. Emerson & James Willis**	***Session #49:*** **PSM Metrics and Culture** Location: Ballroom D **Session Chair: John Wincek**
8:00 am	Oversights and Omissions in Process Hazard Analyses: Lessons Learned from CSB Investigations *(Mark Kaszniak)*	New Statistical Insights into Reasons for Initial Failures in Pressure Relief Valves *(Julia Bukowski)*	Cheddar or Swiss? How Strong are Your Barriers? *(Michael P. Broadribb)*
8:30 am	Optimizing Qualitative Hazard Evaluation *(William (Bill) Bridges)*	Validation of Spring Operated Pressure Relief Valve Time-to-Failure and the Importance of Statistically Supported Maintenance Intervals *(Robert E. Gross and Stephen P. Harris)*	Implementation of Process Safety Management Into Diverse Corporate Cultures *(C. A. Soczek)*
9:00 am	LOPA as Practiced at Pfizer's Global Manufacturing API Facility in Kalamazoo, MI *(G.E Evenson)*	Lessons Learned from Corrosion Under Insulation *(Tim Overton)*	Process Safety Leadership – A Systematic Approach *(C. D. Shelton)*
9:30 am	**Coffee and Networking Break in Exhibit Hall**		

(Tuesday, April 28 continued on next page)

TUESDAY, APRIL 28 Continued

	Session #46: **Hazard Evaluation and LOPA (Continued)** Location: Room 24/25 **Session Co-Chairs: Robert W. Johnson & Mike Moosemiller**	*Session #47:* **Mechanical Integrity — State of the Art (Continued)** Location: Room 22/23 **Session Co-Chairs: Steven D. Emerson & James Willis**	*Session #49:* **PSM Metrics and Culture (Continued)** Location: Ballroom D **Session Chair: John Wincek**
10:15 am	The Buncefield (U.K.) Fire and Explosion: Improving Layer of Protection Analysis Practise to Determine the Required Degree of Protection to Meet Regulator Requirements *(Richard Gowland)*	Implementing Mechanical Integrity at Solutia *(F. Russ Davis)*	Safety Culture: From Black Art to Real Improvements *(Cinzia De Santis)*
10:45 am	The New Approach of Risk Analysis in France for Seveso Establishments: Strengths and Weaknesses *(Jerome Taveau)*	Adoption of Mechanical Integrity Principles at Air Liquide Large Industries US *(Danny Ledbetter)*	OECD Guidance on Safety Performance Indicators *(Kim Jennings and Fran Schulberg)*
11:15 am	Hazard Evaluations and Layer of Protection Analysis: Independent Protection Layers, A New CCPS Guideline Book *(John Murphy)*	Implementing Risk-Based Inspection on a Global Scale *(Andrew Morey)*	CCPS Process Safety Metrics Status Report *(Dorothy Kellog)*
12:00 NOON	**Networking Luncheon**		
	Session #62: **Food and Consumer Products Process Hazards** Location: Room 24/25 **Session Co-Chairs: John E. Going & Nir Keren**	*Session #63:* **Process Safety – From the Top Down to Bottom Up** Location: Room 22/23 **Session Co-Chairs: Philip M. Myers & Donald Abrahamson**	*Session #61:* **Transportation Risk** Location: Ballroom D **Session Chair: Cheryl Grounds**
1:30 pm	Preventing Combustible Dust Explosions in the Food and Consumer Products Industries *(John Bresland)*	The Process Safety Guerilla: Giving Your Company a Future, Even if it Doesn't Want One *(Scott Berger)*	Guidelines for Chemical Transportation, Safety, Security and Risk Management *(Brad Fuller)*
2:00 pm	From the Top Down – Hazard Identification and Risk Evaluation in Food Processing – Toxic Releases, Dust Explosions and Flammable Excursions *(Catherine Kollhoff)*	Using Playbooks – A Unique Strategy in Technical Oversight of Highly Hazardous Operations *(James Gary)*	A Model to Evaluate Pipeline Failure Frequencies based on Design and Operating Conditions *(Vania De Stefani)*
2:30 pm	Nothing to Sneeze at – Combustible Dust Hazard Process Safety *(Guy R. Colonna)*	An Impetus for Management Involvement in Process Safety *(Jack Chosnek)*	Analysis of the Explosion Linked to the Formation of an Explosive Atmosphere In a Shop Carrying Fuel Products *(Sébastien Evanno)*
3:00 pm	**Coffee and Networking Break in Exhibit Hall**		
	Session #62: **Food and Consumer Products Process Hazards (Continued)** Location: Room 24/25 **Session Co-Chairs: John E. Going & Nir Keren**	*Session #63:* **Process Safety – From the Top Down to Bottom Up (Continued)** Location: Room 22/23 **Session Co-Chairs: Philip M. Myers & Donald Abrahamson**	*Session #77:* **Fires, Explosions, and Quantitative Analysis** Location: Ballroom D **Session Chair: Jatin Shah**
3:30 pm	Promoting the Safe Use of Ammonia for Industrial Refrigeration *(Bruce Badger)*	What Does Management Need to Know To Lead Process Safety Improvement? *(Mark Paradies)*	Protection of Permanent Buildings at an Existing Upstream Oil and Gas Facility *(Mark Aspengren)*
4:00 pm	Process Hazards in the Distilled Spirits Industry, Birkmire *(John C. Birkmire)*	Process Safety – Perspectives from the Top *(Tom Archibald)*	Basic Tools for a QRA of an Oilfield Cold Vent *(Pablo Zunana)*
4:30 pm	Replacement of HCFC-22 Refrigerants with Ammonia: a Challenge for Hockey Arenas in Quebec *(Jean Paul Lacoursiere)*		Blast Response of Soft-Skinned Shelters and Occupant Vulnerability *(D.R. Whitehouse)*

WEDNESDAY, APRIL 29

	43rd Annual Loss Prevention Symposium	11th Process Plant Safety Symposiun	24th CCPS International Conference
	Session #82: **Audits, Inspection and Regulation** Location: Room 24/25 **Session Co-Chairs: Lisa Long & Frank H. Gurry**	*Session #83:* **Applications of Risk Analysis Techniques** Location: Room 22/23 **Session Co-Chairs: Alvin Waller & Douglas J. Ferguson**	*Session #79:* **Managing Organizational Change** Location: Ballroom D **Session Chair: Donald J. Connolley**
8:00 am	Auditing Your Process Safety Culture *(David M. Heller)*	Procter & Gamble Global Risk Ranking *(Frank H. Gurry)*	Identifying Key Safety Roles during Organizational Change *(P.A. Davidson)*
8:30 am	Preliminary Findings from OSHA's Refinery National Emphasis Program *(Lisa Long)*	Overfill Protection Systems – Complex Analysis, Simple Solution *(Angela Summers)*	Risk Assessment Process for Organizational Changes *(John C. Wincek)*
9:00 am	Overview of the Chemical Facility Anti-Terrorism Standards *(D. R. Deziel)*	Pressurization of Fixed Roof Storage Tanks Due to External Fire *(Fabien Fouillen)*	An Overview of the CCPS Guideline on Process Safety Consideration During the Merger and Acquisition Process *(G. Melham)*
9:30 pm	**Coffee and Networking Break in Exhibit Hall**		
	Session #82: **Audits, Inspection and Regulation (Continued)** Location: Room 24/25 **Session Co-Chairs: Lisa Long & Frank H. Gurry**	*Session #83:* **Applications of Risk Analysis Techniques (Continued)** Location: Room 22/23 **Session Co-Chairs: Alvin Waller & Douglas J. Ferguson**	*Session #94:* **Risk Criteria** Location: Ballroom D **Session Chair: Karen Tancredi**
10:15 am	Dust Explosion Prevention: Regulations, Standards, and Mitigation Techniques *(Thimothy J. Myers)*	Quantifying the Effectiveness of Explosion Protection Measures *(Peter Moore)*	Single Scenario Risk Criteria: Why Would I Want Them and How do I get Them? *(J. Wayne Chastain)*
10:45 am	Lessons Learned in Auditing Automated Systems for PSM Compliance *(Angela E. Summers)*	The Failure to Communicate: Risk Assessment Challenges Outside the Chemical Process Industries *(Catherine Kollhoff)*	The Application of Peak Risk Concepts within a Corporate Risk Program *(Fred Henselwood)*
11:15 am	A Practical Approach to Implementing a Process Safety Management System Upgrade in Response to an OSHA Refinery PSM NEP Audit *(Robert Walter)*	Are Our Facilities Safe?... And Now?... And Now? *(Melina Alpert)*	Choosing Appropriate Quantitative Safety Risk Criteria: Applications from the New CCPS Guidelines *(Dave Jones)*
12:00 Noon	Luncheon With Speaker: **Lisa Morrison**, BP International		

(Wednesday, April 29 continued on next page)

WEDNESDAY, APRIL 29 Continued

	Session #96: **Case Histories — Joint (CCPS/LPS/PPSS) Session** Location: Ballroom D **Session Chairs: Henry L. Febo, Cheryl Grounds, William Bridges, & Kathy Pearson**
1:30 pm	Lessons Learned from the 2006 Facility Explosion in Danvers, MA *(Scott Davis)*
2:00 pm	Root Cause Analysis of a Cryogenic Refrigeration System Explosion *(Delmar Morrison)*
2:30 pm	Valero – McKee Refinery Propane Release and Fire *(Jim Lay)*
3:00 pm	**Coffee and Networking Break in Exhibit Hall**
3:15 pm	An Oven Explosion Incident – Lessons Learned on PSI, SIS and MOC *(Michael D. Cazabon)*
3:45 pm	Polymerization Reactor Explosion at the XYZ Chemical Company *(Peter Howell)*
4:15 pm	Tank 15 – The Perfect Storm *(Jeffrey Oland)*
4:45 pm	Global Congress Closing Remarks

THURSDAY, APRIL 30

	All papers presented Thursday belong to the 43rd Annual Loss Prevention Symposium (LPS) — Session #111
	43rd Annual Loss Prevention Symposium ***Session #111:*** **Explosion Prevention and Mitigation — LPS Session** Location: Room 24/25 **Session Chairs: Dr. Erdem A. Ural & Brian Kelly**
8:00 am	The Dynamics of Vented Gas Explosions *(C. Proust)*
8:30 am	Safe Thickness for Combustible Dust Layers *(Erdem Ural)*
9:00 am	Explosion Characteristics of Hydrogen-Air Mixtures in a Confined Space *(Dan Crowl)*
9:30 am	**Coffee and Networking Break in Exhibit Hall**
10:15 am	Prevention and Mitigation of Polyethylene and Hydrocarbon/Polyethylene Explosions *(P.R. Amyotte)*
10:45 am	Flammable Gas Venting/Treatment Lessons Learned and Best Practices *(James VanOmmeren)*
11:15 am	Process Management Lessons Learned from a Petroleum Refinery Spent Caustic Explosion *(K.P. Bloch)*
11:45 am	Flammability Characteristics of Low GWP Refrigerant HFO-1234yf *(Barbara Minor)*
12:15 pm	**Session Closing Remarks**

Key Issues with Implementing LOPA (Layer of Protection Analysis) – Perspective from One of the Originators of LOPA

William (Bill) Bridges, Process Improvement Institute, Inc. (PII)
Knoxville, TN, USA
wbridges@piii.com
Tony Clark, Process Improvement Institute, Inc. (PII)
Knoxville, TN, USA

ABSTRACT

This paper focuses on problems observed with LOPA during the first 8-years of broad use.

- One the biggest issues is that organizations use LOPA without following the rules for LOPA, especially the rules related to maintaining, testing, and record keeping for each IPL and for each "optimized" IE. Some companies use the values for an IPL or IE listed in the LOPA guideline book (CCPS, 2001), or elsewhere, but without implementing the management systems to maintain the IPL at the claimed PFD value.
- Another issue is that many companies and analysts overuse LOPA. For instance, IEC 61511 allows a qualitative PHA team to determine if a SIF is needed for a scenario and to specify a SIL 1 or 2, if one is needed. Yet, most folks believe that only LOPA, RiskGraph, or QRA is valid for determining if a SIF is needed and then use the same method to determine what SIL is needed. The LOPA book authors expected the number of scenarios going to LOPA (after a HAZOP/PHA) would be 1% to 10% (max) of those uncovered in a qualitative analysis (maybe after 100 HAZOP nodes, you would do 1-10 LOPA). Some of us believed that a PHA team would recommend (or use) LOPA only if the scenario was too complex for the PHA/HAZOP team. It appears that most companies are using LOPA for every scenario with a severe consequence (greater than 10% of the scenarios).
- Many times there is weak definition of the consequence that is being avoided, so an IPL does not always match up well with the consequence.
- LOPA is also overworked when it is used. Many of us on the original LOPA book authorship considered LOPA a single analyst

job, after a PHA/HAZOP. Instead, the trend appears to be that companies (or perhaps their consultants) make LOPA part of the PHA (in-situ).
- LOPA is used in PHA team settings, which distracts PHA teams from their primary task of brainstorming to identify what accident scenarios can occur.

This paper also summarizes the many benefits LOPA has produced for the industry.

BRIEF HISTORY OF LOPA

The initial development of layer of protection analysis (LOPA) was done internally within individual companies. However, once a method had been developed and refined, several companies published papers describing the driving forces behind their efforts to develop the method, their experience with LOPA, and examples of its use (Bridges, 1997; Dowell, 1997; Ewbank and York, 1997). In particular, the papers and discussion among the attendees at the October 1997 CCPS (Center for Chemical Process Safety, part of AIChE), International Conference and Workshop on Risk Analysis in Process Safety in Atlanta, Georgia, brought agreement that a book describing the LOPA method should be developed.

In parallel with these efforts, discussions took place on the requirements for the design of safety instrumented systems (SIS) to provide the required levels of availability. United States and international standards (IEC, 1998 and 2000) described the architecture and design features of SISs. Informative sections suggested methods to determine the required safety integrity level (SIL), but LOPA was not mentioned until the draft of International Electrotechnical Commission (IEC) 61511, Part 3, which appeared in late 1999. These issues were summarized in the CCPS workshop on the application of ISA S84, held in 2000.

The first LOPA book was developed by a CCPS committee from 1997 through 2000 and was published in 2001. LOPA became widely used following the publication and most companies around the world have used LOPA, with some companies having used LOPA a lot. During roughly 10-years of widespread use of LOPA, and especially during the last 5-years, use of LOPA has greatly accelerated. It is likely that more than 1-million LOPAs have been performed. During this same period, many abuses of LOPA have been noted and several innovations have occurred.

In 2007, CCPS commissioned a new guideline book (1) to expand the list of independent protection layers (IPLs) and initiating events (IEs) and (2) to try to remedy some of the major issues noted in the use of LOPA. The new (future) book is discussed in another paper at this conference; this book is due into publication in early 2010.

INTENT OF LOPA

LOPA is one of many methods for assessing a given scenario to determine if the risk is acceptable. It uses rigid rules to simplify and standardize the definitions of independent protection layers (IPLs) and initiating events (IEs). If these rules are followed, then the simplified risk assessment math of LOPA is valid and the risk assessment should give an order-of-magnitude approximation of the risk of a given cause-consequence pair (scenario). The rules also cover the minimum criteria for maintaining features and task executions that relate to IEs and IPLs.

LOPA is only one option for judging risk. The most common, and still the best method for most scenarios, is the process hazard analysis (PHA) team; their judgment is qualitative, but the "fuzzy" math of the individual team members usually coalesces into excellent judgment of risk for nearly all accident scenarios.

Figure 1, below, updated from the original figure contained in the LOPA book, illustrates where LOPA fits into the spectrum of methods for judging risk.

NOTE: As stated in all books and papers on the topic, LOPA does not find accident scenarios, typically only the qualitative hazard evaluation methods can find new accident scenarios.

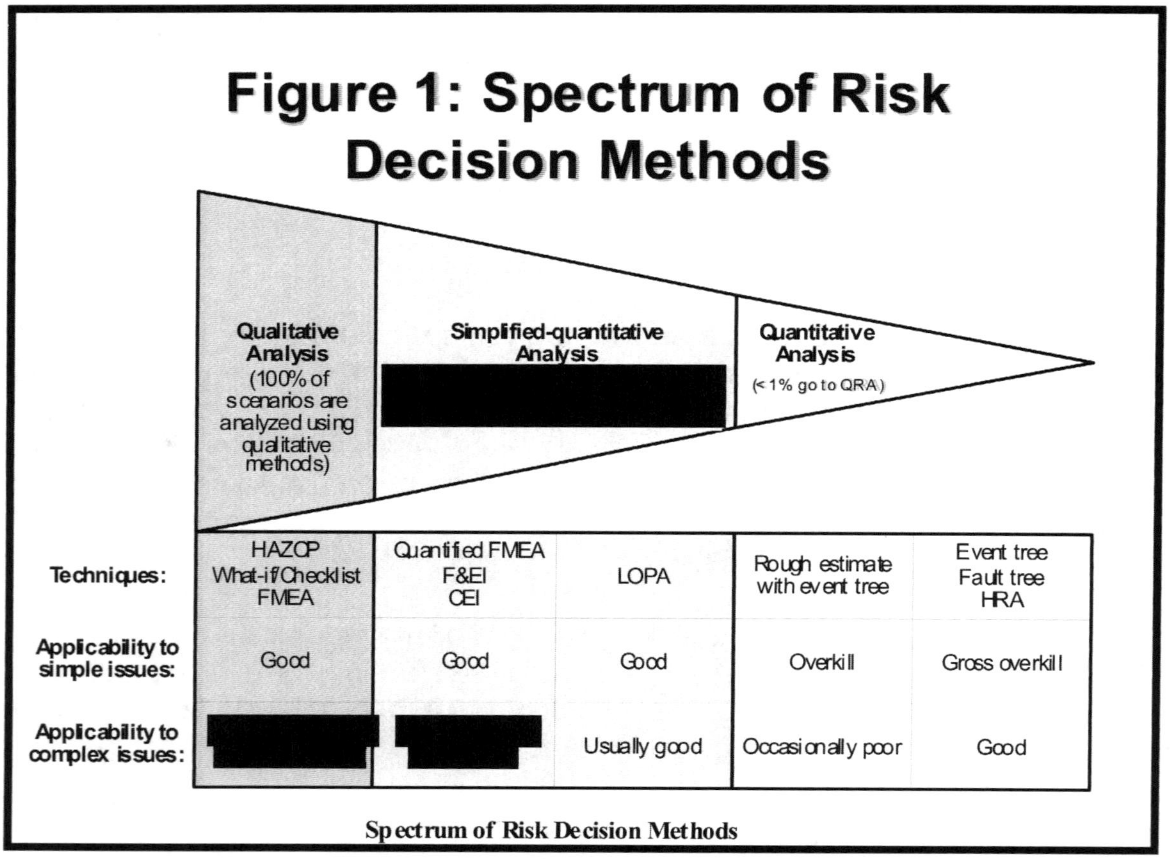

Spectrum of Risk Decision Methods

Courtesy of Process Improvement Institute, Inc., 2004.

RELATIONSHIP TO SIL DETERMINATION

LOPA started with and continues to have a unique relationship with SIS and SIL determination. Some of the originators of LOPA needed LOPA to defend against an arbitrary assignment of safety instrumented functions (SIFs) for systems that were already "adequately" safeguarded by other means. This became apparent in the mid-1990s with the early development of SIS standards within chemical companies and by the Instrument Society of America (ISA). Some of these early standards would have imposed a minimum SIL for a given consequence, without much regard for the number and value of other IPLs that already existed or were viable alternatives to the SISs. Much of these arbitrary requirements for SIS have disappeared, but some remain.

For the most part today, LOPA is seen as one tool (in many parts of the world, the preferred tool) for determining if an SIF is necessary and if it is the correct choice of risk reduction and for determining what SIL is necessary, if an SIF is chosen as the risk reduction method. With that said, PHA teams are also allowed by IEC 61508 to make these same determinations.

However, some organizations use LOPA to answer the question: "What SIL is needed to lower the risk to the risk target?" without first asking, "Are we at tolerable risk already?" or "Are there better alternatives for lowering the risk?" This leads to a huge over-specification of SIFs (and the wasting of resources to design, implement, and maintain these SIFs) and to many spurious shutdowns of units (which also waste money and increase the risk of accidents that can occur during re-start of the process).

SUMMARY OF GAINS FROM LOPA

Overall, LOPA has been highly successful in helping organizations to understand and control risk. Below is a listing of some of the benefits derived from LOPA:

- More consistent definition for protection layers. Information from LOPA helps an organization decide which safeguards to focus on during operation, maintenance, and related training. For instance, many companies decide to focus their inspection, test, and preventive maintenance activities on the IPLs identified during LOPA; these companies often decide to run the remaining safeguards (those not identified as IPLs) to failure or subject them to less rigorous test and maintenance schedules. Therefore, LOPA is a tool for implementing a comprehensive process safety management (PSM) mechanical integrity (MI) or risk-based maintenance system, and it aids in the identification of "safety critical" features and tasks.

- LOPA helps identify operations and practices that were previously thought to have sufficient safeguards, but on more detailed analysis (facilitated by LOPA), the analysis indicates there are far more safeguards than necessary. Therefore, the company/site can eliminate unnecessary safeguards (eliminate "extra" IPLs) or perhaps similarly eliminate the care and proof of these safeguards (i.e., run the extra IPLs to failure) – while still maintaining tolerable risk.

- Improved evaluation in PHAs of which "listed" safeguards are "valid" safeguards. Even though LOPA was not meant for "live" use during a qualitative, team hazard review (because it normally distracts a team from their primary job of brainstorming what can go wrong), one of the main benefits realized from the first decade of implementing LOPA within an organization is that the qualitative hazard reviews (such as PHAs or HAZOPs) have improved. This is because now those in a qualitative hazard review are not likely to list a protective feature or action as a safeguard in a HAZOP or What-if based analysis, unless that safeguard appears to meet (or is intended to meet) the requirements of an IPL (discussed later). This has improved the qualitative risk judgment of such teams because the safeguards they list and discuss are very likely independent, capable, validated, and audited.

- Justifying when SIS is not needed (proper SIL determination can usually show no SIL is needed). LOPA helps provide the basis for a clear, functional specification for an IPL (ANSI/ISA 84.00.01 [2004] and IEC 61508/61511 [1998; 2001; 2004]) and also facilitates assigning of the SIL needed (a quantitative approach is required by these standards for a SIL of 3).

- Faster quantification of severe risk scenarios. LOPA requires less time than quantitative risk analysis (QRA). This benefit applies particularly to scenarios that are too complex for qualitative assessment of risk. A full QRA of a scenario can take hours or days, whereas as a LOPA for the same scenario will likely take about 1-hour.

- Common language among risk reviewers around the world. LOPA helps resolve conflicts in decision making by providing a consistent, simplified framework for estimating the risk of a scenario and provides a common language for discussing risk. LOPA provides a better risk decision basis compared to subjective risk judgment, especially for complex scenarios. Related to this point, LOPA gives clarity in the reasoning process and it documents everything that was considered. While this method uses numbers, judgment and experience are not excluded. In some cases, the LOPA analyst (or in some cases, LOPA team) was uncomfortable with the number calculated, so the analyst reviewed the assumptions for the frequency of the initiating event.

- LOPA offers a rational basis for managing layers of protection that may be taken out of service (such as bypass of a shutdown interlock). LOPA also provides clarity on what to pay attention to when an IPL is out of service for testing or repair.

SUMMARY OF ISSUES WITH THE CURRENT IMPLEMENTATION OF LOPA

While LOPA has been a great benefit to industry, we have observed many issues with the implementation of LOPA over the past 10-years of use.

1. **One of the biggest problems with LOPA is that its users do not always follow the rules of LOPA.** IPL and IE values are picked from a list, while the specific IEs and IPLs are not (1) validated to have the stated value and (2) not maintained to sustain the stated value. Below is a listing of the rules for IPLs (with impact on IEs as well), and descriptions of the problems we have observed:

- **IPLs must meet independence rule.** This most important rule is ***not*** often violated, at least not intentionally. But it is violated occasionally. For instance, a LOPA may use two basic process control system (BPCS) loops without first verifying that the BPCS qualifies for Approach B, as outlined in Chapter 11 of the LOPA guideline (CCPS, 2001). Or, sometimes the LOPA will re-use an operator or use another operator within the same work team; this usually will not pass the test of independence. Part of the reason for this latter problem is the lack of clarity in the first LOPA guideline. The next book being drafted by CCPS (scheduled for release in 2010) for IPLs and IEs will provide more clarification on the use of human IPLs. The basic rule is that you cannot use any work group (like an operations shift or maintenance/operator team doing online maintenance activities) more than once.

- **IPL and IE values must be defensible.** This has been a problem. Many organizations choose values from handbooks (or from the original LOPA book) and papers/articles or obtain them from calculations based on discrete component failure rates from databases, and then assume those values apply to their situation. This is not a good assumption. The overriding factor in the reliability of a component, or the reliability of the human action is often the local control of human error and the local environment of the equipment operators. For example, a PSV in clean, gas service has a much different reliability than a PSV in olefin or acid service.

- **IPLs and IEs must be maintained such that they produce the IE or IPL values stated.** This has been a huge problem in the past 8-years of LOPA implementation and is one of the problems we hope to fix with the new CCPS textbook on IPLs and IEs (pending 2010). An IPL cannot be assigned any risk reduction value if it is not maintained well enough to produce the risk reduction value. Part of the problem is that the industry is still struggling to know what tasks and how much effort (frequency) is needed to get these values. This is partly because the consensus codes and standards (except for the SIS standards) were developed *without* a specific IPL value in mind. So, organizations must maintain their IPLs and causes of IEs in a way that gives the probability of failure on demand (PFD) that they use in LOPA. Where does an organization find this information on best practices for maintaining critical systems? Consensus codes provide a starting point for many IPLs and IEs; we expect these to gradually improve and they should eventually provide the anticipated PFDs (or failure rates) if the practices are followed. Plant data should be looked at to make sure the IE or IPL is not "outside" of the bounds expected. In the interim, have very experienced operations and maintenance staff on the PHA teams (where scenarios are first identified and where the raw input data for LOPA is identified) and also have these same staff provide the maintenance practices, test practices, and operator drill routines for use within an organization.

- **IPLs and IEs must be validated and records must be kept and audited.** This has been a huge problem in the past 8-years of LOPA implementation and is one of the problems we hope to fix with the new CCPS textbook on IPLs and IEs (pending 2010). Currently, even if we follow industry advice, it means nothing if our own test data shows the IPL or IE value is something less than what we wanted. For instance, what if you follow industry advice for PSV maintenance and testing, but then your own records indicate that every time you pull a couple of specific PSVs, they are compromised in some way? Obviously, you have a problem with these specific PSVs and, therefore, using them as IPLs (or using the IPL value you hoped for) is not valid.

  Part of the problem is that the industry is still weak on reporting near misses. For many of us, anytime we have challenged the last IPL or two, and anytime we find an IPL in a failed state, we have a near miss. Yet are these being reported and investigated? In most cases, they are not. There should be 20-100 near misses reported for each loss event, yet the ratio in the industry is currently 1-2 (Bridges, 2000 and 2008). The organization that gets many near misses reported (and a large percentage of these also get investigated), will have tremendous gains in loss prevention and will also have a much better idea of their reliability factors supporting the PFD values for IPLs and IEs failure rates.

  Most companies we deal with recognize they must have an inspection, test, or PM program for component and instrumented IPLs. But, most companies do not have a test program for response of humans to critical alarms or similar indications. Human action must be validated and documented to be an IPL. The specificity and frequency of such testing is still under debate, but it needs to occur.

  To follow is an example of the upcoming CCPS guideline on IPLs and IEs, which illustrates the requirements necessary for an organization to claim a given IPL and stated PFD value.

Example of Criteria Necessary to Qualify an IPL (extracted from rough draft of pending IPL/IE guideline, expected early 2010)

IPL Description	IPL PFD	Consequence Prevented	Special Considerations, including Special Independence Conditions	Proof Method	Proof Frequency
Deflagration Flame Arrester or Stable Detonation Arrester installed inline between an ignition source (e.g., TOX) and a source of flammable or combustible vapors	0.1	Deflagration, not detonation	- The piping between the ignition source and arrestor is well below the run-up distance required to allow a transition to detonation (DDT) for Deflagration type or formation of Unstable Detonation for a Stable Detonation type. - Location considered (avoid "hot side" on bottom of vertically mounted arrester since this decreases endurance burn; avoid accumulation of liquids in arrester, use drains to remove liquids). - Device does not impose excessive flow restriction on the process and any fouling issues have been addressed. - Temperature monitoring with a thermocouple directly in contact with the hot side of the device is highly recommended to allow operations to recognize when device is being challenged. Reference "Deflagration and Detonation Flame Arresters", Stanley S. Grossel, CCPS, 2002.	- Device is included on a routine maintenance schedule which specifies shutting down the line and opening the device for inspection. - Device is always inspected if it is suspected to have stopped a flame or if process upset could compromise its integrity. - Inspection includes determining whether the device is plugged and whether corrosion might compromise its capability to arrest a flame in accordance with most industry standards.	Initially, every 12 months or per vendor recommendation, then adjust the interval to 24, 36, or up to a maximum of 4 years, if no signs of corrosion.
Fire suppression system (water; water and foam; other suppressants); automatic	0.1	Explosion or further consequence beyond the external pool fire which can arise for whatever causes (such to prevent BLEVE of vessels, runaway reactions resulting in internal detonations, etc.). Not applicable to jet fires and not applicable to prevent a forming vapor cloud from exploding.	Using fire or smoke detectors to automatically activate a system designed to prevent or control a fire. For example, wet pipe or dry pipe systems with fusible links, deluge systems or water curtains applying water, foam. Can be within pipe or duct for flame-front suppression. Can be externals sprays for external fires (such as deluge system under pipe racks or under tansl/columns). See NFPA 13 Installation of Sprinkler Systems, NFPA 15 Standard for Water Spray Fixed Systems for Fire Protection, NFPA 16 Standard for the Installation of Foam-Water Sprinkler and Foam-Water Spray Systems or NFPA 750 Standard on Water Mist Fire Protection Systems for additional detail.	See NFPA 25 Inspection, Testing and Maintenance of Water-Based Fire Protection Systems for additional detail on what to test, how to test (methods), for water supply (tank, pond, etc.), water pressure source (head tank, pumps), activation systems, etc.	See NFPA 25 Inspection, Testing and Maintenance of Water-Based Fire Protection Systems for additional detail on testing frequency.
Fire suppression system (non-aqueous including dry agent) for room; automatic	0.01	Fire within an Enclosure	Refer to NFPA 17 Dry Chemical Extinguishing Systems, NFPA 2001 Clean Agent Fire Extinguishing Systems or NFPA 12A Standard on Halon 1301 Fire Extinguishing Systems for information on Dry Powder, Clean Agent and Halon Flooding Systems.	Refer to NFPA 17 Dry Chemical Extinguishing Systems, NFPA 2001 Clean Agent Fire Extinguishing Systems or NFPA 12A Standard on Halon 1301 Fire Extinguishing Systems for information on Dry Powder, Clean Agent and Halon Flooding Systems for Inspection, Testing and Maintenance Requirements.	Refer to NFPA 17 Dry Chemical Extinguishing Systems, NFPA 2001 Clean Agent Fire Extinguishing Systems or NFPA 12A Standard on Halon 1301 Fire Extinguishing Systems for information on Dry Powder, Clean Agent and Halon Flooding Systems for Inspection, Testing and Maintenance Requirements.

- **Many times there is weak definition of the consequence that is being avoided, so an IPL does not always matchup well with the consequence.** This can cause both over- and under-estimates of the risk.

  One issue that we have come across is that the worst case consequences are being assumed for failure of a control system, which sounds wise, but for some cases, it is **overly pessimistic**. For instance a full bore pipework rupture is assumed due to brittle failure if the pipework is subjected to temperatures lower than its design temperature. While catastrophic brittle failure is remotely possible (this may only occur in 1 in 50 or 1 in 100 cases), we'd get a much better indication of the risk if operators recorded each occasion and the consequences of exceeding design parameters, even if nothing happened. Otherwise, we believe that we are being too pessimistic.

  Similarly, for overpressure scenarios, we see LOPA teams stating that the consequence will be catastrophic loss of containment if the pressure exceeds the set point of the PSV, whereas we normally have to reach 150% of the set point before we start to see leaks and normally have to teach >200% of the set point before the releases are large. And this 200% is still well below the deformation pressure for the system. On the other hand, the committee writing the new CCPS book on IPLs and IEs has been convinced by industry data that the PFD for a PSV is likely at 0.01 instead of 0.001, as stated in the example table of the first LOPA book (CCPS, 2001).

  > ***In the collective experience, there has been a tendency to overestimate the risk causing companies to spend far too much money on safety systems that are not necessary.***

  That said, there are cases where the **risks have been underestimated**, caused by predicting the consequences to be less severe than they would be. One illustrative example of this, is the Buncefield UK Incident (Buncefield, 2008), in which overfilling of one of the petrol tanks resulted in a series of explosions, which caused a huge fire, engulfing 20 large storage tanks (the largest fire in the UK since World-War II). The fire burned for 5 days. No one was killed, but there were 43 minor injuries. The incident happened early on a Sunday morning, but had it occurred during a normal working day there could have been a significant number of fatalities. The economic impact added to around £1 billion ($1.45 billion USD), which accounted for the emergency response, compensation for loss, costs to the aviation sector and the investigation.

  > Consider conducting a LOPA on overfilling of a petrol tank before the incident. For the consequences, most LOPA analyst would have assumed that the petrol would run down the sides of tank and collect as a liquid in the bund (dike), which it did. But on igniting, what would you have

assumed, bearing in mind that the area was not particularly confined? A pool fire in the bund (dike), most likely; serious, but not catastrophic. Few analysts would have perceived such massive explosions since the understanding was that petrol does not easily explode. The consequences, and hence the risk would therefore have been under-estimated and IPLs we consider necessary today would have been deemed over-kill.

2. **Overuse of LOPA.** Though we love LOPA, many of the originators thought LOPA would be used a lot less frequently than it is now. As shown in Figure 1, it was anticipated that LOPA would be used on 1-5% of the scenarios uncovered in PHAs. Further, though a couple of companies at the time were using LOPA within the PHA setting, it was anticipated that LOPA would eventually be used "after" a PHA team meeting. The various examples of overuse are discussed below:

   - **Use within PHAs.** Many of us on the original LOPA book authorship considered LOPA a single analyst job, after a PHA/HAZOP, for just a few scenarios (maybe after 100 HAZOP nodes, you would do 1-10 LOPA). Instead, the trend appears to be that companies (or perhaps their consultants) make LOPA part of the PHA (in-situ). If the PHA/HAZOP team is properly disciplined on what qualifies as a safeguard (a qualitative definition of an IPL from LOPA), then performing LOPA in situ is overkill. In most situations, a qualitative team (HAZOP team) can make just as good or better judgment than provided by LOPA. LOPA is just another way to make a decision, has many pitfalls, and doesn't work for many types of scenarios.

     - **Distracts PHA team from brainstorming** – The PHA teams have an awesome responsibility which is to find ALL scenarios that can lead to adverse risk, and do this for all modes of operation. This is the organization's one opportunity for finding most of these scenarios. Near misses and accidents are the other ways of finding these scenarios, but the industry (on average) still has trouble getting near misses reported, so our options for finding risky scenarios before catastrophe strikes is thin. Further, the mind is in a different pattern when brainstorming, compared to when analyzing. Analyzing is usually easier, so we tend to drift that way naturally. But, the job of the diverse members of a PHA team is to brainstorm. Since it is difficult to bring the team back into brainstorm mode after switching to analyzing mode, it is best for an organization to limit the amount brain-power expended by PHA teams on analysis.

       One of the initial reasons for developing LOPA was to move semi-quantification of risk **out** of the PHA sessions, to enhance brainstorming. Notice the title that Art Dowell put on the first paper he wrote on the topic: "LOPA: **After** PHA/HAZOP, Before QRA" [emphasis added]. Now, many organizations are making the same mistake as in the early 1990s and moving LOPA (semi-quantification of risk) back into the PHAs. If an organization allows this, they will distract their multi-disciplinary teams from brainstorming and then they will miss accident scenarios. It makes no sense to analyze known scenarios in much more detail than before, while not finding other scenarios. It is the "undiscovered" scenarios that will cost the company more in the long run, since we will not have proper safeguards for many of these undiscovered scenarios.

- **Takes time away from what is critical for the PHA team to do: Identify scenarios for ALL modes of operation.** LOPA is taking time away from proper identification of risky scenarios. For instance, folks feel now that they ***must*** do LOPA to validate the risk decision of PHA teams. However, those same PHA teams are NOT given time to do a proper identification of scenarios for all modes of operation. Many (perhaps most) PHA teams focus 90% on normal mode of operation, instead of focusing mostly on startup, shutdown, and other abnormal operations. Hence, they are not finding the scenarios that account for 70% of the accidents that occur. Yet, the same managers and staff who support LOPA for every scenario (doubling the time of an efficiently led qualitative risk review and risk judgment) are the same ones who say "What-if or 2-guideword analysis of startup and shutdown procedures takes too long" (yet this expansion of the qualitative studies to evaluate all modes of operation only takes about 50% more PHA team time). Perhaps in the future we will agree that more LOPA after a PHA will be beneficial, currently most PHA teams are only finding one half of the accident scenarios (focusing nearly entirely on normal [typically continuous] modes of operation). We realize this is not a problem with LOPA itself, but to us it is an issue related to LOPA and is similar to the past trend in the UK where QRA was pushed hard (and still is) for each facility, even though they have not completed a proper PHA of all modes of operation. The only difference between the over-emphasis on QRA and what is happening now is that LOPA produces a faster QRA. The industry is still missing many of the largest and least protected scenarios in favor of over-analyzing the scenarios they know about. Chapter 9 of the HEP Guideline, 3rd edition (CCPS, 2008) explains how to analyze non-routine modes of operation.

- **Use for every Medium and High Risk Scenario**. This issue is likely covered adequately in the discussion above, but increasing the number of scenarios that must go through LOPA, reduces the resources available to find the undiscovered scenarios and to manage existing layers of protection (both issues are listed above). To follow is the guide we use to decide when a LOPA is required (Category 5 is equivalent to consequences greater than $100,000,000 and/or with potential multiple fatalities):

When to use LOPA ***(courtesy of Process Improvement Institute, Inc., 2004)***

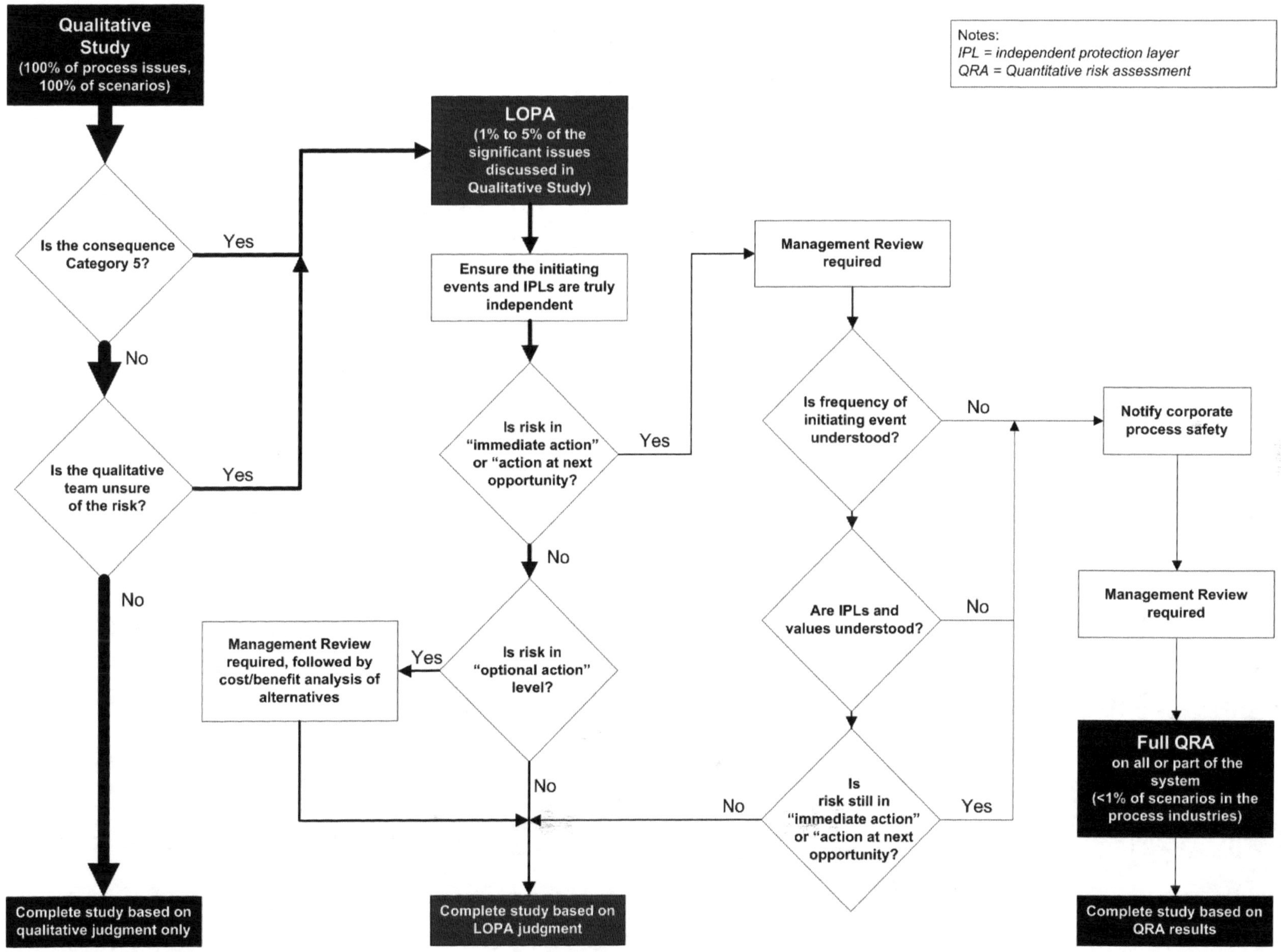

- **Use in studies that are redundant to PHAs, such as "separate SIL determination."** IEC 61511 allows a qualitative PHA team to determine if a SIL is needed for a scenario and to specify a SIL 1 or 2, if one is needed. Yet, most folks believe that only LOPA or RiskGraph or QRA is valid for determining if a SIL is needed and then use the same methods to determine what SIL is needed. As a result, many people do LOPA on almost every scenario of moderate consequence or higher. The LOPA book authors expected the number of scenarios going to LOPA (after a HAZOP/PHA) be 1% to 10% (max) of those uncovered in a qualitative analysis, and some of us believed that usually a team would use LOPA only if the scenario was too complex for the PHA/HAZOP team. SIS standards allow a PHA team to determine (1) when a SIF is Not required and (2) what SIL is needed if an SIF is required (though for SIL 3 and higher, a LOPA or similar study is recommended by SIS standards).

3. **Too many resources dedicated to LOPA studies.**

   - **Typically, one LOPA analyst is sufficient (if he/she has easy access to experts within the organization).** Once a LOPA is completed for a scenario, the results can be relayed to management or to a PHA team, or similar decision makers. The mention of a LOPA team in the first LOPA book was anecdotal, but many organizations now require a LOPA team (instead of single analyst). Some companies used a LOPA team early because (1) the analyst trained in LOPA was not in the PHA session, so translation from the PHA team to the analyst was necessary in many cases and (2) LOPA was new, so more heads were needed to decide "Is this the right way to apply LOPA." However, if the LOPA analyst was on the PHA team or if the teams get used to communicating to the LOPA analyst(s), then one person can perform the LOPA. Note that no brainstorming is necessary for LOPA, so there is no inherent need for a team.

   - **Why use a LOPA team (with a LOPA leader and LOPA scribe)?** There is almost No brainstorming occurring during a true LOPA analysis so there is no need for a team. On the other hand, if the LOPA team (or PHA team) recommends a SIL, then a small team (2-3 experts) will be needed to "specify the SIL design and functionality issues (such as sequence and delays) for the SIL". Also, later someone (usually one person) will be needed to validate the SIL design will produce the SIL determined by the PHA or LOPA team.

4. **Too much emphasis on software.**

   - **You do not need software for a 1+1+2=4 calculation (i.e.,** "Why use a sledgehammer to crack a nut?"). Most of the commercial packages for documenting PHAs (using HAZOP, What-If, or whatever methods) have options for sending scenarios to LOPA worksheets. These can ease the completion of LOPA exporting of some data for LOPA; in fact, I designed one of the first such applications for LEADER software. On the other hand, these PHA software options do not make it easier to document "why an IPL is valid." Many analysts and most operating companies have implemented their own spreadsheet applications, which:

- Take very little to develop
- Are easy for others in the company to learn
- Can be linked to internal reliability data tables for company-approved IPLs and IEs
- Are easy to use on multiple work-stations
- Are easy to add and edit text that describes the scenario and factors
- Are often easier to use than PHA software

The most important need of LOPA documentation is to type the scenario description in detail, explain clearly why an IPL is given credit, and most importantly, describe how each IPL is maintained to sustain the credit given. This can all be done freehand, and PHA (or LOPA) software does not help shortcut this necessary chore.

Closing

Introduction of the streamlined quantitative risk assessment method, LOPA, has had a tremendous impact on the chemical and related industries. 90% of the quantitative risk assessments that may be necessary can now be performed in 1/10th the time of a QRA. Many benefits have been reaped, including a continual improvement on the identification and control of critical features and actions. However, the initial rollout of LOPA has led to a few problems, as discussed in this paper. The problems are easily remedied by judicious use of LOPA and by carefully adhering to the rules of LOPA.

References

1. *Guidelines for Hazard Evaluation Procedures, Third Edition with Worked Examples*, CCPS/AIChE, New York, NY, 2008.
2. *Initiating Events and Independent Protection Layers for LOPA, CCPS/*AIChE, New York, NY, 2010 [pending].
3. *Layer of Protection Analysis (LOPA); Simplified Process Risk Analysis*, CCPS/AIChE, New York, NY, 2001.
4. Bridges, W.G. (with T.R. Williams), "Risk Acceptance Criteria and Risk Judgment Tools Applied Worldwide Within a Chemical Company," *International Conference and Workshop on Risk Analysis in Process Safety, CCPS/AICHE*, Atlanta, GA, October 1997.

5. Bridges, W.G., "Getting Near Misses Reported - Revisited," *8th ASSE-Middle East Chapter Conference and Workshop*, Bahrain, February, 2008.

6. Bridges, W.G., "Get Near Misses Reported," *International Conference and Workshop on Process Industry Incidents, CCPS/AICHE*, Orlando, FL, October 2000.

7. Dowell, A. M., III, "Layer of Protection Analysis: A New PHA Tool, After Hazop, Before Fault Tree," *International Conference and Workshop on Risk Analysis in Process Safety*, CCPS/AIChE, Atlanta, GA, October 1997.

8. Ewbank, Rodger M., and Gary S. York, "Rhône-Poulenc Inc. Process Hazard Analysis and Risk Assessment Methodology," *International Conference and Workshop on Risk Analysis in Process Safety, CCPS/AIChE*, Atlanta, GA, October 1997.

9. IEC (1998), IEC 61508, *Functional Safety of Electrical / Electronic / Programmable Electronic Safety-related Systems, Parts 1–7*, Geneva: International Electrotechnical Commission.

10. IEC (2001), IEC 61511, Functional Safety Instrumented Systems for the Process Industry Sector, Parts 1–3, Geneva: International Electrotechnical Commission.

11. The Buncefield Incident, 11 December 2005, The Final Report of the Major Incident Investigation Board, Volume 1 (2008).

A Large Layer Of Protection Analysis For A Gas Terminal 2000+ Scenarios/'Cause – Consequence Pairs'

Richard Gowland, European Process Safety Centre

- The scope of the study was a large gas terminal handling North Sea gas, condensates etc.,
- Supplying into Northern Europe Gas distribution
- Separation of LPGs and gasolines for shipment and sale.

This was a large upgrade project engineered by a major U.S. contractor on a site where the existing Protective Systems include:

- High Integrity Pressure Protection Systems (HIPPS)
- Process Shut Downs (PSDs) – plant control trips from control instrumentation
- Manual Emergency Shut down Systems (ESDs)
- Pressure relief to flare and atmosphere.

The object of the study was to answer the 2 questions: For hazard scenarios

- is there adequate prevention and protection to meet company and national requirements.
- If more protection is required, what form should it take and if Safety Instrumented Systems are required, what Safety Integrity Level (SIL) is needed?

Layer of Protection Analysis (LOPA) is a method which examines a scenario from its initiating events and their frequency, through all modifying probabilities, probability of failure of protective systems by simple calculation to see if a tolerable frequency of the final top event meets or exceeds requirements.

Setting the consequence severity tolerability levels:

A key requirement of the LOPA methodology is to have severity criteria and target tolerability criteria agreed (injury/environmental/financial). This had been done by the owners and were generally in agreement with industry and regulator commonly adopted data for injury and fatality criteria. Company criteria were used throughout the study. An example of this was the adoption of a a frequency of 1E-05 for an on site fatality. It was agreed that the LOPA study team would define the severity for each scenario and case. This would then be assigned the standard target tolerated frequency to start the study. A software tool based on Microsoft Excel ® was produced to calculate any protection gaps and record the studies as they proceeded. This had embedded in it all target frequencies, failure rates, conditional modifiers (probability of ignition, probability of exposure etc.) to ensure consistency and transparency. This would evolve into a single standard and tool for whole company.

- Wherever possible, we created standard consequence:
    - Injury/fatality
    - Cost
    - Environmental effect
- Agreed to always use worst case and carry out single 'line' study for each deviation (cause/consequence pair)

Initiating events - Failure frequencies:

- Most were agreed at the outset and included in the software. These were 'process deviation causing events' such as process control loop element failure, loss of containment and human error. Some data used was 'generic industry taken from literature and some supported by company data

•

There was a concentration on process control deviations and equipment failures

- Level/flow/pressure/temperature control
- Heat exchanger tube failures
- Unplanned trips (turbines and compressors)

And some human error cases, e.g.

- Manually controlled systems
    - Tank level management
    - Loading/unloading
    - 'Pigging'

Modifying probabilities/Conditional Modifiers:

Probability of ignition.

Hydrocarbon releases were assumed to ignite if their full potential was to be studied. Large releases were assumed to ignite 100% of the time. This was the default position when doubt was sufficient. Other smaller releases were assigned a probability of ignition based on physical properties, quantities, phase etc.

Probability of Exposure:

Was based on operational and geographical factors. For example a risk in a tank farm which is visited rarely and only for inspection would be considered for 1% or 10% probability of exposure. For very large events, again the default was always 100%.

In practice, during the study, conditional modifiers were rarely applied.

Protective Systems:

Probability of Failure on Demand for Independent Protection Layers/Barriers:

- The following were normally considered:
  - Control systems
  - Process Shutdowns (PSDs)
  - High Integrity Pressure Protection Systems (HIPPS)
  - Others (relief etc.)
- Wherever necessary Reference was made to IEC 61511, NL 'Purple Book' and National Gas Industry implementation standard.

As the study went on we created standard set of scenarios for each unit operation. The basic list of these operations is:

- **Loading/Unloading**
- **Storage**
- **Pressure let down stations**
- **Pigging stations (launching and retrieving)**
- **Mole sieves/Driers**
- **Heaters/re-boilers/coolers**
- **Distillation towers**
- **Reflux drums**
- **Pumps**
- **Compressors**
- **Turbines**
- **Knock out drums/flares**
- **Boilers**

The STUDY

Loading/Unloading scenarios/cases:

Primarily loading out (LPGs and Hydrocarbon liquids)

Failures on:

- level control
- Pumps
- Loading arms
  - Loading arm failure frequencies not readily available
  - Needed to estimate from company experience and extrapolating NL Purple Book data
  - Ship movement control
  - Human error

Discipline was needed on assumption about time at risk based on number of loading activities per year.

Storage: These are large storage tanks, cryogenic and ambient.

Storage System scenarios/cases: It was considered that since level is managed by operations the study primarily went into into overfill and overpressure cases.

- Refrigerated storage for LPGs.

  - ❖ Typical Initiators:
    - Loss of level indication
    - Wrong valve line up
    - Operator error

  - ❖ Independent Protection Layers (IPLs) and their Probability of Failure on Demand (PFD)
    - Level alarms 1e-01
    - Inflow trips via block valves 1e-01
    - Vent control to flare** 1e-01**
    - PSV Relief to flare 1e-02
    - PSV Relief to atmosphere 1e-02**
    - Position switches on valves 1E-01

  ** Care needed – not always accepted. Capacity may be limited or discharge not safe.

- Atmospheric storage (usually Floating roof) for flammable condensates etc.
  - IPLs
    - Level alarms 1e-01
    - Inflow trips via block valves 1e-01
    - PSV Relief to flare 1e-02
    - ***PSV Relief to atmosphere 1e-02 ****

  ***** Advised against this as IPL***

Fortunately – LPG stores were adequately protected without considering atmospheric relief as Independent Layer of Protection. The same was not true for atmospheric storage where it was found that relief would occur to tank roofs and lead to similar hazards to those seen at Buncefield with flammable liquids cascading down the sides of the tank.

'Pressure Let down' stations:

Gas from offshore pipelines in at 160 bar and needed to be distributed into the gas terminal and processing units at lower pressures. This was achieved by conventional pressure control valves acting on signals from analogue pressure transmitters. In most cases the design pressure of the downstream equipment was much lower than the maximum available pressure and therefore pressure protection is required. Conventional relief systems are not appropriate. The pressure protection loops of interest usually comprised pressure switches connected to a logic solver and outputs to large fast acting block valves.

- Initiators – primarily failure of control valve loop.

- Independent Layers of Protection:
    - Extensive use of High Integrity Pressure Protection Systems (HIPPS) which give tight shut off and containment and in a few cases relief to flare
    - HIPPS function met requirements of Norwegian Gas industry and IEC 61511 requirements – PFD of 1E-02 normally used
    - Hybrid systems sometimes used (so called secondary HIPPS) – incorporating control valve as final element. Changes were needed to allow testing – also name change. Reluctant to assign a value of better than 1E-01 to PFD.
    - Valve Closure time on very large lines was an issue which eliminated some protection

The HIPPS system was already evaluated by reputable consultant and the principal findings supported LOPA team decisions.

Pigging Operations: The network of pipelines entering and exiting the terminal is routinely inspected and cleaned by the use of 'pigs'. The launching and retrieval activity for these necessitates opening the pipeline at 'pigging stations'. This activity is regarded as hazardous since it has the potential for large gas releases.

- Extensive protective systems
    - Limit switches
    - Mechanical locks
    - Block valves

  are used.

Since the equipment is very specialised and most actions are manual, most data was taken from experience and engineering judgement. The outcome was a strengthening of procedures to reduce failure frequencies.

Mole Sieves and Driers: These operations were primarily for quality control, however, accelerated corrosion could occur if the system failed.

- Operational Mode Failure consequences slow to build up and readily detected by quality testing
- Key hazard potential is in the regeneration process
  - Reactivity – exotherms
  - Overheating
  - Overpressure
- Initiating Events usually easy to define (control system failure)

Heaters/Reboilers/Coolers:

- Used industry or NL 'Purple Book' failure frequencies
- Exchanger design and pressure protection prevented system rupture
- Most hazard caused by high pressure gas into utilities - Almost always a protection gap resulted
- Outcome emphasised
  - Mechanical integrity needs to reduce failure frequencies by eliminating known failure phenomena. Need to limit extension of inspection intervals justified by Risk Based Inspection (RBI)
  - Action on final release points and leak detection to reduce potential consequences – not true Independent Protection Layers

Distillation:

- Very Large capacity trains
- Level and pressure control on feed forward from bottoms of each distillation column to inlet of next column in the train
- Potential to overpressure downstream equipment if level control is lost
- Level Control normally from displacement level controllers
- Pressure and Level scenarios more dominant than Temperature and Flow
- Independent Protection Layers
  - Alarms and trips from Pressure transmitters (independent of level control)
  - HIPPS
  - Multiple PSVs (PSV data checked on line during study – exploited high degree of design data availability – leading to re-evaluation of relief system)
  - Some concerns on independence – plant operators in the team were very quick to discover problems

Reflux Drums and Condensate Receivers:

- Level control deviations provided most scenarios which could over-pressurise the system
  - High pressure controlled by vent control to flare or in extreme circumstances, PSV to flare
  - Overflow
- Controls and trips generally shown to be adequate
- Level control deviations provided most scenarios
  - High pressure controlled by vent control to flare or in extreme circumstances, PSV to flare
  - Overflow
- Controls and trips generally shown to be adequate

Pumps:

- Key issues were:
  - Operation with closed valves
  - Seal leaks
    - Single
    - Double
    - Alarms
  - What are the severities of consequences?
    - Determined by size, potential ignition and location
    -

Compressors:

Hazards examined included leakage, catastrophic failures as a result of liquids in gas feeds.

- Mechanical and electronic machine protection devices to trip on vibration etc.
- Hazard cases protected by:
  - Liquid separation drums on inlet
  - Machine protection (vibration etc.) (not used as hazard protection in most cases)
- Surge conditions controlled by anti surge controls
- Settle out pressure upstream caused concern on compressor trips
- Concerns about machine protection being disabled by 'sharing' normal control system elements (Common Cause Failure re-design needed)
- Mechanical and electronic machine protection devices to trip on vibration etc.

Turbines:

Hazards examined included leakage, catastrophic failures as a result of liquids in gas feeds.

- Hazard cases protected by:
  - Machine protection (vibration etc.) (not used as hazard protection in most cases)
- Surge conditions controlled by anti surge controls
- Pressure ratings normally eliminated overpressure scenarios
- Concerns about machine protection being disabled by 'sharing' normal control system elements (Common Cause Failure re-design needed)

Knock out drums and Flares:

In these cases, the hazards studied were availability of flare in relief cases. Multiple relief requirements, liquid removal.

- All flares have knock out drums
- Knock out drums normally maintained empty
- Reliefs (PSVs) and pressure let downs (PCVs) discharge to knock out drums via headers
- All knock out drums have multiple alarms from independent field sensors
- High liquid level starts pump or opens drain – not usually taken as IPL
- Need high reliability operator response to level deviations since operator can be seen as a common cause of failure
- Serious consequences in fuel gas systems if liquid carryover
- Other ideas for trips/SISs?

Boilers:

Boilers are fired primarily on fuel gas from the terminal itself.

- Most boiler controls (Burner Management Systems) needed to be SIL2
- Serious concerns about fuel gas with liquid entrainment or gross quantities of liquid carry over into combustion chambers
- Some issues on hydrocarbons in condensate (detection and response)

Results and conclusions from the study.

- In all, there were more than 2000 cases studied by a team which was staffed by a leader, process engineers, instrument designer and operators. The LOPA methodology proved to be robust and was readily understood by all members. The plant operators quickly became familiar with the technique and were able to verify and in some cases dispute operational practices and the functional modes of plant operation. Their contribution proved to be vital. The use of simple excel based software was helpful since it ensured understanding and consistency of approach. There were occasional problems deriving failure frequencies and probability of failure of some unusual hardware systems, but this difficulty would apply to any method and not simply to LOPA. There were also some procedural issues around requests from instrumentation designers to assign a Safety Integrity Level (SIL) rating to every control or 'trip' loop in the facility. A) Managing Contractor Instrument department expected to start from a list of instruments – then to assign each loop with a SIL rating

Specifically, Plant Operations and LOPA chair expected to start with the deviations/consequences/scenarios – wherever possible using HAZOP as inputs

- Starting with deviations/consequences/scenarios
    - Addresses only those concerns which are significant
    - Connects to previous HAZOP studies
    - Cross check with Cause and Effect diagrams from instrumentation design to make sure no process related hazards missed
- Starting from a list of instrument loops
    - Large number of loops involved in process and quality control.
    - Process control items automatically included

This second approach was demonstrated to be unnecessary if the studies addressed all the hazardous scenarios and then include those instrument systems which were relevant.

Major lessons from the study were:

- In most cases, adequate protection appeared to be provided, but in a significant number of cases failed the independence or adequacy/effectiveness requirements
- Closing times for very large valves presented significant challenges.
- Management and procedures needed to be reinforced, documented and training given on tank level management and response to alarms
- The operating company's equipment records (design capacities etc.) were excellent and helped determine the adequacy of protection against studied hazards.
- Mechanical Integrity programmes for heat exchangers were very important in reducing failure frequencies, since there were few true independent protection layers available and protection was mainly in the role of consequence alert and reduction (e.g. gas detection)

Improved Integration of LOPA with HAZOP Analyses

Dick Baum
Nancy Faulk
John Pérez, P.E.
Lloyd's Register Celerity3, Inc.
Houston, TX 77024

ABSTRACT

Integrating Layer of Protection Analysis (LOPA) with Hazard and Operability Analysis (HAZOP) has many advantages over performing these studies separately. The merits include: fewer actions from the combined effort compared to performing only a HAZOP; team continuity resulting from the combined effort as opposed to two separate teams having possibly differing points of view; and, ultimately, a time and cost savings realized by the combination. This integration defines the risk associated with a given scenario, enabling better decisions which impact business assurance.

By using the Center for Chemical Process Safety (CCPS) guidelines to define the independent protection layers upfront, the gray areas can often be reduced or eliminated; thereby enabling a more thorough LOPA. Examples include taking credit if a unit has two independent operators (outside and inside) responding to critical alarms, or taking credit for centralized control rooms which may allow immediate operator interaction and response.

This paper shows how the guidelines have been used successfully in past joint HAZOP/LOPA studies, and describes initial preparation protocol that can ensure high-quality results.

INTRODUCTION

Process safety management (PSM) has been in place in many companies around the world since 1992, after government mandates in Europe, the U.S, and elsewhere came into effect. Recently, the Center for Chemical Process Safety (CCPS) published guidelines for risk-based process safety (RBPS) as a framework for the process industries to implement and maintain PSM activities *(1)*. These guidelines include twenty RBPS elements designed to improve process safety and effectiveness, of which one element is Hazard Identification and Risk Analysis (HIRA). HIRA comprises all activities involved in addressing the main risk questions of hazard, consequence and likelihood, and includes a spectrum of tools; which tool is chosen depends on the level of complexity of the scenario and may require multiple review processes. The following are done in order, proceeding as far down the list as is necessary to provide sufficient analysis without waste of resources:

1. The team initially performs simple hazard identification or qualitative risk assessment, through hazard and operability analysis (HAZOP), what-if/checklist analysis, or failure modes and effects analysis (FMEA).
2. If the scenario is too complex or the consequences too severe to rely solely on the qualitative assessment, the team moves to semi-quantitative risk analysis, utilizing layers of protection analysis (LOPA), or failure modes, effects, and criticality analysis (FMECA).
3. The decision to proceed to a quantitative technique, such as fault tree analysis or chemical process quantitative risk analysis (CPQRA), occurs, e.g. when the team determines the scenario is too complex or critical to rely on LOPA for risk assessment. It is estimated that 1% of scenarios proceed to this stage *(2)*.

The focus of this paper is the common practice of applying LOPA after an unacceptable consequence of a credible scenario are identified in the HAZOP, and will not consider any cause that requires a more rigorous quantitative technique.

HOW LOPA IS USED

LOPA is a semi-quantitative tool for understanding, through a rational, objective, risk-based approach, how many safeguards are needed for a particular accident scenario, and subsequently for performing calculations to determine whether existing protection is adequate. The CCPS responded to industry requests to tabulate and present industry practice for LOPA, and published its landmark book on the subject in 2001 *(2)*. Since that time, LOPA has become a widely accepted tool for risk analysis and for safety integrity level (SIL) assignment. The quantitative calculation of risk is based on estimated frequency of potential incidents and the probability of failure on demand (PFOD) of the protection layers *(2, 3)*. The steps are outlined in the literature, for those unfamiliar with the procedure *(2)*.

TIMING OF THE LOPA PROCESS

Early on in its utilization, the LOPA process was commonly viewed as an independent study, to be facilitated at a later date than the HAZOP. Usually separate teams performed the different analyses. For example, the LOPA team might have consisted of a process engineer and the LOPA team leader, or even a single analyst, in the case where the scenarios have been previously determined by a hazard evaluation team; in other cases, a larger team or an independent LOPA analyst reviewed the work *(2, 4)*. This separation was thought to be useful because *(5, 6)*:

- concern that teams would become burned out;
- software was not available that converted HAZOP cause/consequence into a LOPA scenario;
- fear of inflating the cost of the proceedings; and
- prevention of biases from the HAZOP team.

It became apparent that many of these concerns were overstated *(6, 7)*; in addition, some problems arose because of the separation of the analyses. In many instances, facilitators were discovering that the LOPA team members wanted to recreate the HAZOP because they disagreed about the scenarios which had been developed. Completed HAZOP study records cannot be changed or modified without re-convening the original HAZOP team and performing a review of the questionable scenarios. This requirement for a HAZOP review has the potential to frustrate new LOPA members, if they disagree with the previous results.

In contrast to the separate HAZOP/LOPA analyses discussed in the previous paragraphs, some of the original companies performing LOPA began incorporating it as part of their HAZOP sessions. Their positive experiences led to discussions about the timing of the LOPA process; this evolved into an issue discussed in the early CCPS book *(2)*. Those authors suggested that LOPA could be integrated with the HAZOP, or subsequent to the HAZOP, and either process would be effective. From that point in time until now, there have been lively discussions concerning this topic *(5-8)*, with both sides asserting that their methodology is superior. This paper shows how the guidelines have been used successfully in past joint HAZOP/LOPA studies, and describes initial preparation protocol that can ensure high-quality results.

RECENT TRENDS

In recent years, the trend of integrating LOPA with HAZOP has grown substantially *(6-8)*; those facilitators who were integrating the two analyses identified many benefits:

- **Expertise –** HAZOP participants are usually experts; getting a similarly qualified group together at another time is difficult and costly;

- **Reduced time** – Keeping the same team in play reduces time lost in getting a new team up to speed on the systems involved. Even if the same team is utilized, memories dim and some review and ramp-up time are necessary;
- **Ease of integration** – Several software tools are now available that maintain a common database in order to link causes, consequences and safeguards; and
- **Cost savings** – Estimates of up to 30% savings were noticed as much as four years ago *(7)*.

The process industries are requesting integrated LOPA with HAZOP in response to these associated benefits. The data collected by the Risk Management Services division of Lloyd's Register Celerity3, for example, shows the following trend in the percentage of clients requesting the integrated approach when executing their HAZOPs:

- 2006 – 10%
- 2007 – 25%
- 2008 – 80%

The dramatic rise is an indication of the widespread recognition that LOPA is a growing trend, and that there are advantages in this integrated approach. What follows is a discussion of ways facilities can enhance their success.

IMPROVED HAZOP WITH LOPA ANALYSES

This discussion will focus on three areas where experience shows the most potential for improving the integrated LOPA with HAZOP process: management decisions, training, and tracking changes.

(1) Management Decisions

Several key decisions must be made in advance of initiating the HIRA. LOPA is usually applied to determine if the calculated risk from a scenario is within the risk tolerance criteria or if the risk must be reduced. Thus, the first decision which must be established is the choice of criteria protocol to use. The most commonly used risk judgment protocols are: the use of a risk matrix; occasionally a numerical criteria method; and rarely (generally not recommended) the use of expert judgment *(2)*.

This paper assumes the use of the risk matrix, where each cell of the risk matrix correlates to the degree of risk reduction required for a cause-consequence pair in that cell, and is based on the consequence frequency and severity. The number of cells is in turn based on the number of rows and columns predetermined by management; will the matrix be a 4x4 or a 5x7? How many of the cells will trigger LOPA, based on management's tolerance of risk? Many companies have found that once they establish the matrix and the trigger points, the objectivity of the risk categories enables risk decisions to be made faster and more consistently. Proper calibration of and establishment of the trigger points in the risk matrix are vital to ensuring neither over- nor under-conservative specification are made *(8)*.

The next management policy decision is providing well-defined IPLs, or Integrated Protection Layers. An IPL is a device, system or action that is capable of preventing a scenario from proceeding to its undesired consequence, and must meet the following requirements *(2)*:

- Effective in preventing the undesired consequence, and quantifiable in terms of its PFOD
- Independent of the components or actions of any other IPL already claimed for the same scenario;
- Independent of the occurrence of consequences of the initiating event; and
- Auditable, by testing, documentation or review, in order to verify the assumed effectiveness.

While the requirements seem straightforward, there are many instances where safeguards are confused with IPLs or where IPLs are not clearly defined. A safeguard is defined similarly to an IPL: any device, system or action that would likely interrupt the chain of events following an initiating event. The distinction is that its effectiveness cannot be quantified, due to lack of data or uncertainty as to independence *(2)*, which leads to the commonly quoted assertion that all IPLs are safeguards, but not all safeguards are IPLs. The management or the LOPA analyst must determine if a safeguard meets the requirements listed above, and subsequently assign a PFOD based on either literature values or industry experience. The CCPS has provided guidelines that provide a baseline for IPL determination, both active and passive *(2)*. In addition, the reference provides a range of PFODs. Another source of guidelines for PFODs is the International Electrotechnical Commission *(9)*. It has been the authors' experience that the more effort management puts forth providing a listing of acceptable IPLs and specifying the PFODs to use, the less gray area exists for the LOPA team. This has the benefit of providing more objectivity, of speeding up the LOPA process, and of ensuring that the analysis follows management guidelines and not the team's interpretation.

(2) Assure Proper Training

Another improvement issue discussed here is the resolve to use competent personnel, with the appropriate training. Consider developing and then using HIRA Subject Matter Experts (SMEs). These may be the process SMEs (such as operations, process chemists, reliability engineers, rotating equipment specialists), who are called upon to be part of the team. There should also be a few SMEs who are trained to *lead* HIRA activities routinely, including HAZOP/LOPA. Additionally, others with assigned HAZOP/LOPA duties would benefit from additional training, e.g. scribe or even participant. Note, however, that the team leader should normally come from outside the facility, independent of the operation *(1, 2)*.

The risk program should have dedicated personnel for managing HIRA activities, scheduling, tracking resolutions and so forth. There may be training involved for these personnel also. The decisions that management makes come in the form of support of the risk program personnel and support for appropriate training.

(3) Track Changes

Every facility should have implemented a system for auditing the elements which were identified as IPLs, either through testing or inspection, and provide maintenance or replacement as necessary. There should exist a listing of the IPLs, including design criteria, limitations, PFODs, and initiating causes for which they mitigate.

Suppose, as a result of a LOPA or an audit, a change to IPL definitions or PFOD values, or a recommendation to add an IPL, occurs. These actions must be tracked as an item for review by the organization's safety committee:

- If after review a change in definition of IPL or PFOD values is accepted, this change needs to be updated in the appropriate published set of specifications, and the IPL summary sheet needs to be updated.
- Any recommendation for additional IPLs must be evaluated in order to reduce risk; implementation should be included in updates.

This ensures that at the next LOPA, the analyst and team have the current IPL information and are able to apply definitions consistently.

A HAZOP/LOPA can be thought of as an audit for the facility's risk management program and provides a method of continuous improvement. Items with unacceptable risk or those procedures which conflict with management guidelines are identified and assigned to appropriate personnel for resolution, as action items. Most HAZOPs produce a list of items in a node generally termed the "Parking Lot" or a "Comment Page" that records items needing further evaluation and study, but are not necessarily entered into an action tracking database. Management must ensure follow-up on these concerns as well, and track all action items through to resolution.

ADDITIONAL EXAMPLES

This section of the paper will discuss unresolved situations observed by the authors when facilitating integrated HAZOP with LOPA sessions. Involvement of key personnel early on will ensure that the right questions are asked, and answered, in order to establish the appropriate technical risk matrix for that facility. Significant investments of time and resources made at the project front-end result in lower time and energy expenditures and less subjectivity.

Example 1: What PFOD will the facility use for its relief valves? Literature suggests a range from $10^{-1} - 10^{-5}$ *(2)*, and the PFOD must be determined based on that facility's inspection, testing and maintenance program (are the valves checked once a year or every five years?).

Example 2: Whether or not to use the BPCS (Basic Process Control System) as an IPL is a concern, as it is a relatively weak IPL with limited testing capability. Human error, e.g. in bypassing alarms, can increase the probability of failure. Care must be taken to ensure that the BPCS is not associated with the initiating event. IEC 61511 limits the combined PFOD to ≥ 10^{-1} for any BPCS IPL that can be applied *(9)*.

Example 3: Some facilities will accept a low priority alarm as an IPL, but other facilities do not. Operator action, initiated by an alarm, is also accepted by some facilities, which have assessed the effectiveness of that operator action and determined that it meets the criteria required for an IPL. Human IPLs are generally less reliable than engineering controls, but not crediting human actions – assuming adequate documentation, training and testing – is too conservative.

Example 4: Is it possible to count multiple functions in one BPCS as separate IPLs for a given scenario? Recall that one of the requirements for an IPL is that it is independent of the components of any other IPL, but in some situations it is possible. This requires independence of sensors and final elements, as well as adequate analysis and documentation.

Example 5: If a unit that has multiple operators which can be notified of an unsafe condition; can these each count as an IPL? Consider the case of having an inside operator and an outside operator; in order to count them both as IPLs, there must exist independent alarms or sensors which would signal operator action independently.

Example 6: The size of the risk matrix must be determined, and the LOPA matrix needs to mirror the HAZOP matrix. In one instance, the authors found that a facility used different matrices for HAZOP and LOPA, which caused the LOPA team to have to re-evaluate the consequences. This could have been easily prevented with better alignment of the risk matrix philosophies.

CONCLUDING STATEMENTS

LOPA is a semi-quantitative risk analysis technique, triggered when the qualitative assessment of a HAZOP shows the scenario to be too complex or the consequences too severe. Facilities are increasingly integrating their HAZOP analyses with their LOPA studies, primarily because of the many benefits: a qualified group is already convened; study time is reduced; common databases can link causes, consequences and safeguards; and cost savings are realized. This paper has presented three areas where experience shows the most potential for improving the integrated LOPA with HAZOP process: management decisions, training, and tracking changes. In addition, several examples were presented on the types of decisions which must be agreed upon, in order to ensure objectivity as well as to save time and effort when performing these analyses. The integrated approach of LOPA with HAZOP can be a vital part of a successful and robust HIRA program.

References

1. Center for Chemical Process Safety (CCPS), "Guidelines for Risk-Based Process Safety," American Institute for Chemical Engineers, New York, NY (2008).

2. Center for Chemical Process Safety (CCPS), "Layer of Protection Analysis: Simplified Process Risk Assessment," American Institute for Chemical Engineers, New York, NY (2001).

3. Summers, A. E., "Introduction to Layers of Protection Analysis," J. Hazard. Mater., 104 (1-3), pp. 163-168 (2003).

4. Dowell, A. M. and D. C. Hendershot, "Simplified Risk Analysis – Layer of Protection Analysis (LOPA)," AIChE 2002 National Meeting, Indianapolis, IN, Nov. 3-8, 2002.

5. Marszal, E., "Dispelling Myths about SIL Selection," http://www.isa.org/InTechTemplate.cfm?Section=Standards_Update1&template =/ContentManagement/ContentDisplay.cfm&ContentID=58275, International Society of Automation, Research Triangle Park, NC.

6. Dowell, A. M., and T. R. Williams, "Layer of Protection Analysis: Generating Scenarios Automatically from HAZOP Data," Process Saf. Prog., 24 (1), pp.38-44 (2005).

7. Bingham, K. and P. Goteti, "Integrating HAZOP and SIL/LOPA Analysis: Best Practice Recommendations," ISA 2004, Houston, TX, October 5-7, 2004.

8. Kallambettu, J., "Adopting Safety Life Cycle Activities for SIS by Engineering, Procurement, and Construction Companies," Safety Control Systems Conference 2008, Calgary AB, May 28-29 2008.

9. International Electrotechnical Commission (IEC), IEC 61511 "Functional Safety: Safety-Instrumented Systems for the Process Industry Sector, Part 1," Geneva, Switzerland (2003).

The Impact of the BP Baker Report and Process Safety Survey

Stephanie C. Payne, Ph.D.
Department of Psychology
Texas A&M University
4235 TAMU
College Station, TX 77843-4235
(979) 845-2090
spayne@psych.tamu.edu

Jennifer M. Rodriguez
Department of Psychology
Texas A&M University
4235 TAMU
College Station, TX 77843-4235

Mindy E. Bergman, Ph.D.
Department of Psychology
Texas A&M University
4235 TAMU
College Station, TX 77843-4235

Support for this research was provided by the Mary Kay O'Connor Process Safety Center.

ABSTRACT

In March 2005, the BP Texas City refinery experienced an explosion that killed 15 and injured 170 employees. As a result of this catastrophic event, the U.S. Chemical Safety and Hazard Investigation Board (CSB) recommended that an independent panel assess BP North America's safety management, corporate safety oversights, and corporate safety culture specific to process (i.e., focusing on the safety of process activity, rather than on occupational safety behaviors of personnel). Thus, the BP U.S. Refineries Independent Safety Review Panel was formed. The Panel conducted an extensive review including interviews, refinery visits, survey administration, and relevant document review in an attempt to respond to the CSB's concerns and later released their findings in the BP Baker Panel Report (Baker, 2007). The current study was designed to gather information from industry insiders—safety personnel and chemical engineers—about the impact of the BP Baker Report on personal and process safety within their respective organizations and the industry as a whole. It was also designed to determine to what extent the BP Process Safety Culture Survey has become a benchmarking tool for assessing process safety culture.

Anecdotal evidence suggests that the BP Baker report has awakened the industry to the importance of differentiating between personal and process safety (M. S. Mannan, personal communication, May 1, 2008; Hopkins, 2009). Among the major findings of the BP Baker Panel was the conclusion that while corporate leaders of BP North America had set a positive tone regarding *personal* safety, they had not done so with *process* safety. The BP Baker Panel concluded that BP North America attended to personal safety indicators (e.g., injuries, days missed from work) and erroneously inferred positive results for personal safety to indicate good process safety (Baker, 2007). In fact, the Panel found that BP's personal injury rates were not predictive of process safety performance. Indicators of personal safety are not indicators of process safety, thus they are not substitutable (CSB, 2007).

In the current study, we administered a survey to 379 subscribers of the Mary Kay O'Connor Process Safety Center listserv in February of 2009. The survey gathered information on the extent to which the BP Baker Panel Report has been read by and advocated by engineers in the field, the extent to which BP Baker Report has had an impact on or been useful within the industry, as well as whether instruments used in the BP Baker Report, such as the BP Process Safety Culture Survey, are also being used by others in the field. Descriptive results from the study will provide information that will be useful in determining to what extent the BP Process Safety Culture Survey may be a useful benchmarking tool from which to gage organizational process safety. It will also provide some insight into the mental models of chemical processing personnel regarding the transportability and generalizability of lessons learned in one organization to others.

INTRODUCTION

In March 2005, the BP Texas City refinery experienced an explosion that killed 15 and injured 170 employees. As a result of this catastrophic event, the U.S. Chemical Safety and Hazard Investigation Board (CSB) recommended that an independent panel assess BP North America's safety management, corporate safety oversights, and corporate safety culture specific to process (i.e., focusing on the safety of process activity, rather than on occupational safety behaviors of personnel). Thus, the BP U.S. Refineries Independent Safety Review Panel was formed. The Panel conducted an extensive review including interviews, refinery visits, survey administration, and relevant document review in an attempt to respond to the CSB's concerns and later released their findings in the BP Baker Panel Report (Baker, 2007). The current study was designed to gather information from industry insiders—safety personnel and chemical engineers—about the impact of the BP Baker Report on personal and process safety within their respective organizations and the industry as a whole. It was also designed to determine to what extent the BP Process Safety Culture Survey has become a benchmarking tool for assessing process safety culture.

Anecdotal evidence suggests that the BP Baker report has awakened the industry to the importance of differentiating between personal and process safety (M. S. Mannan, personal communication, May 1, 2008; Hopkins, 2009). Hopkins (2009) noted *personal safety* refers to safety hazards that affect the individual but are not necessarily affected by the activities of the plant, whereas *process safety* is defined as hazards that arise directly from processing activity at a plant. In high reliability organizations, such as chemical processing, nuclear power plants, and airlines, the importance of understanding and monitoring process safety indicators is very high, as failures of the process can be catastrophic (EPA, 1989; Kletz, 1993).

Among the major findings of the BP Baker Panel was the conclusion that while corporate leaders of BP North America had set a positive tone regarding *personal* safety, they had not done so with *process* safety. The BP Baker Panel concluded that BP North America attended to personal safety indicators (e.g., injuries, days missed from work) and erroneously inferred positive results for personal safety to indicate good process safety (Baker, 2007). In fact, the Panel found that BP's personal injury rates were not predictive of process safety performance. Indicators of personal safety are not indicators of process safety, thus they are not substitutable (CSB, 2007).

The current study is a first look at changes in an essential and hazardous industry—chemical processing—as organizations within the industry undergo cultural transformations regarding safety. Although chemical processing has long been concerned with hazard reduction (i.e., solving engineering problems so that processes run with less risk to personnel and the organization), anecdotal evidence suggests that the BP Baker report is a paradigm shifting document, recommending chemical processing companies attend more closely to their cultures and to understand and monitor both personal and process safety as distinct entities in the organization. Our study begins to examine the extent to which consensus is coalescing around the BP Safety Culture Survey as a possible benchmarking tool in the industry. Should this happen, it would allow organizational researchers to develop industry-wide tools to aid in process safety and process safety culture improvement through the design of training, process monitoring tools, and benchmarking instruments.

METHOD

In February of 2009, approximately 5000 subscribers of the Texas A&M University Mary Kay O'Connor Process Safety Center listserv were invited to participate in an online survey about the impact of the BP Baker Report and Process Safety Survey on the process safety industry. Three hundred seventy-nine surveys were returned for a response rate of approximately 7.6%. Subscribers were given three weeks to complete the 15-min survey and reminded about the survey one week before it closed. They were also offered the opportunity to be included in a raffle for one of five $50 gift cards in exchange for their participation in the survey.

Our sample consisted of 293 males (77%), 38 females (10%), and 48 individuals who chose not to indicate sex. Respondents ages ranged from 21 to 83 with an average of 51.18 years (*SD* = 10.78). The ethnic breakdown of our sample was as follows: 266 (70.1%) Caucasians, three (<1%) African Americans, 16 (4.2%) Hispanics, four (1.0%) Native Americans, thirty-two (8.4%) Asians, two (<1%) Middle Easterners, and 17 (4.4%) designated other. When asked about their current position, 90 (23.7%) indicated they were engineers, 60 (15.8%) indicated they were consultants, 55 (14.5%) indicated they were safety managers or directors, 37 (9.8%) indicated they were safety personnel, 25 (6.6%) indicated they were operations managers, 24 (6.3%) indicated they were process safety personnel or managers, 16 (4.2%) indicated they were government employees, 49 (12.9%) did not provide this information, and the remaining 6.2% indicated some other position (e.g., student, faculty, retired, etc.) On average, respondents had 23.94 (*SD* = 11.45) years of industry experience, mostly in the chemical processing industry.

In addition to asking general demographic questions, we assessed the extent to which respondents were affiliated with BP or worked near Texas City. Ninety-five (25.1%) respondents were affiliated with BP in some way (work/worked for them directly, contract/contracted with them,), and 197 (52%) respondents indicated they had friends or colleagues that were affiliated with BP. Further, 53 (14%) respondents indicated that they work near (within 50 miles of) Texas City, TX now, and 84 (22.2%) participants indicated they had worked near Texas City in the past.

RESULTS AND DISCUSSION

In an attempt to assess participants' familiarity with the BP Baker Report, respondents were asked to what extent they had read the full BP Baker Report. The majority of respondents (86.3%) indicated they had read at least some portion of the report (52 [13.7%] respondents indicated they had not read any of the report.). Of those, 44 (11.6%) indicated they had read a little bit, 130 (34.3%) indicated they had read some, 106 (28%) indicated they had read all, and another 47 (12.4%) indicated they had read the report more than one time.

When asked further about their experiences with the report (check all that apply), 67 (17.7%) indicated they were required to read the report as a part of their job, 50 (13.2%) indicated their supervisor asked their opinion of it, 102 (26.9%) indicated they presented information about the report to others in their company, 248 (65.4%) were personally interested in the information contained in the report, and only two (0.5%) respondents indicated receiving continuing education credit(s) for reading the report.

We assessed the impact of the BP Baker Report through a series of questions that were responded to on a 5-point Likert scale. First, we asked how important the report is to the future of safety in chemical processing. On average, respondents felt it was important (M = 4.12, SD = 0.86) with 240 (63.3%) respondents indicating "important" or "extremely important." Second, we asked if they agreed with the findings from the report. On average, respondents tended to agree with the findings (M = 4.21, SD = 0.76) with 265 (69.9%) respondents choosing "strongly agree" or "agree." Third, we asked to what extent the report was useful to the respondent and his/her organization. On average, respondents reported it was useful (M = 3.97, SD = 0.91) with 226 (59.6%) individuals checking "useful" or "extremely useful." Finally, we asked to what extent the respondents felt the findings in the report generalized to other chemical processing plants. On average, respondents thought the majority of the findings apply to other plants (M = 3.95, SD = 1.01) with 204 (53.8%) respondents selecting "about 75% of the findings apply to other plants" or "all of the findings apply to other plants."

As noted earlier, one of the things highlighted in the BP Baker Report is the distinction between process (or workplace) and personal (or worker) safety. Survey respondents were asked to what extent they were aware of this distinction prior to the BP incident and dissemination of the report. Most of the respondents (208, 54.9%) indicated that they were aware of this distinction to a great extent. We also asked how related respondents believed process and personal safety were to one another. The majority of respondents (130, 34.3%) indicated "one cannot exist without the other" implying a perfect relationship between process and personal safety, and another 122 (32.2%) respondents indicated "improvements in one lead to improvements in the other" implying a positive relationship between the two. Interestingly, 38 (10%) respondents felt "process and personal safety are not related to one another." Finally, we asked to what extent the process and personal safety distinction influences the way respondents deal with safety. On average, respondents indicated this distinction influenced their actions at least to some extent ($M = 2.0$, $SD = 1.26$) with 70 (18.5%) respondents choosing "to some extent" and another 146 (38.5%) respondents indicating "to a great extent."

The next set of questions concerned the BP 2006 Process Safety Culture Survey which was included in the BP Baker Report as an Appendix. First, participants were asked if their company had conducted any surveys about safety-related issues prior to the BP-Baker Report coming out in January of 2007. Most respondents (169; 44.6%) indicated conducting at least one survey prior to January of 2007; however, 70 (18.5%) indicated they had not conducted any surveys. When asked about administering the BP 2006 Process Safety Culture Survey questions, 89 (23.5%) respondents indicated that they had not administered these questions; however an almost equal number of respondents (84; 22.1%) indicated that they had administered BP questions in some capacity.

In summary, the BP Baker Report is generally familiar, important and generalizable to industry insiders and their organizations. Most people agreed with the findings in the report and found the information useful to them and their organizations. While most people were aware of the distinction between process and personal safety and most people believe the two are positively related to each other, there are still some individuals who believe they are not related to one another. Further, this distinction appears to have some but not an extensive amount of influence on the way safety is dealt with in organizations.

Whereas a good number of organizations have conducted safety-related employee surveys, the content of these is likely to be quite variable as less than 25% of the responders have administered BP process safety questions in some capacity. Perhaps not enough time has transpired for the BP safety questions to emerge as a benchmarking tool from which to gage organizational process safety. Perhaps of greater concern is the number of organizations that had not conducted any employee surveys which could identify problems within their own organizations.

We continue to analyze these data and additional survey results will be reported at the conference in April.

REFERENCE

1. The BP US Refineries Independent Safety Review Panel. (2007). The Report of *the BP US Refineries Independent Safety Review Panel.* From http://www.bp.com/liveassets/bp_internet/globalbp/globalbp_uk_english/SP/STAGING/local_assets/assets/pdfs/Baker_panel_report.pdf.

2. Hopkins, Andrew (2009). Thinking about process safety indicators. *Safety Science, 47,* pp. 460-465.

3. Kletz, Trevor (1993). *Lessons from disaster: How organizations have no memory and accidents recur.* Houston, TX: Gulf Publishing.

4. U.S. Chemical Safety and Hazard Investigation Board. (2007). *U.S. Chemical Safety and Hazard Investigation Board Report.* From http://www.csb.gov/completed_investigations/docs/BPFinalReport.pdf.

5. U.S. Environmental Protection Agency (1989). *Accidental release information program.* Washington, DC: EPA.

Identifying "Culture Issues" Using Root Cause Analysis

Mark Paradies
System Improvements, Inc.
238 S. Peters Road, Ste 301
Knoxville, TN 37923
http://www.taproot.com
865-539-2139

ABSTRACT

Culture – assessing and changing it – is a hot topic in many industries. Often the topic is made so complex that only a "culture PhD" could understand the assessment methods. However, culture issues can be identified using already existing root cause analysis data and a simple culture model that can help any root cause analyst or manager take the first step to understanding and changing an organization's culture.

WHAT IS "SAFETY CULTURE"

Culture: It's "the way we do things around here."

Definitions of Safety Culture come from a variety of investigation committees and safety oversight/regulatory agencies:

Columbia Accident Investigation Board[1]:

> "Organizational culture refers to the basic values, norms, beliefs, and practices that characterize the functioning of a particular institution. At the most basic level, organizational culture defines the assumptions that employees make as they carry out their work; it defines 'the way we do things here.' An organization's culture is a powerful force that persists through reorganizations and the departure of key personnel."

UK Health and Safety Commission[2]:

> "The safety culture of an organization is the product of individual group values, attitudes, perceptions, competencies and patterns of behaviour that determine the commitment to, and the style and proficiency of, an organization's health and safety management."

International Atomic Energy Agency, International Nuclear Safety Advisory Group[3]:

> "That assembly of characteristics and attitudes in organizations and individuals which establishes that, as an overriding priority, nuclear plant safety issues receive the attention warranted by their significance."

US Nuclear Regulatory Commission[4]:

> Safety Culture is, ``The necessary full attention to safety matters," and the ``Personal dedication and accountability of all individuals engaged in any activity which has a bearing on the safety of nuclear power plants. A strong safety culture is one that has a strong safety-first focus."

Institute of Nuclear Power Operations[5]:

> Safety Culture is, "An organization's values and behaviors - modeled by its leaders and internalized by its members - that serve to make nuclear safety the overriding priority."

If all these organizations don't see eye-to-eye on the definition of Safety Culture, how can we analyze and fix it? Certainly, these complex definitions must be accompanied by complex analysis systems and equally complex improvement initiatives.

DOES ASSESSING SAFETY CULTURE NEED TO BE COMPLEX?

When dealing with problems - especially Safety Culture problems - regulators and researchers tend to see complexity everywhere. This results in complex issues that are hard for managers and line workers to understand and fix. Lack of true understanding of the issues results in delays in taking action or perhaps taking action that is ineffective.

But does assessing Safety Culture need to be complex? What if we could define some fairly easy to understand cultural models that could be diagnosed with a fairly easy to use root cause analysis tool? This might make cultural assessments and corrective actions much more approachable for management and employees.

SIMPLE DESCRIPTIONS OF SAFETY CULTURE ISSUES

The first step is to define some fairly simple Safety Culture models. This paper will explain seven Safety Culture models that were first presented to industry personnel at the 2008 TapRooT® Summit[6]. They are:

- Start-Up Culture
- Git-er-Dun Culture
- See-No-Evil Culture
- No One Told Me Culture
- Poor Problem Solving Culture
- Backlog Culture
- Bad Safety Culture

The ***Start-Up Culture*** is seen in many small, new companies (start-ups). There aren't many, if any, written rules. People often use their judgment when confronted with options because no one has created a standard approach to work. This culture may be seen even at major facilities when they hire small contractors to perform work.

The ***Git-er-Dun Culture*** can best be described as production above all else. Short-cuts are often taken and rules may be bent or ignored to get work done (as long as no one is looking). Rewards are given for accomplishing tasks. Managers may hear supervisors say: "Boss – you really don't want to know how we did it."

The ***See-No-Evil Culture*** is closely linked to the Git-er-Dun Culture. In the See-No-Evil Culture there will be a lack of effective audits and assessments so that management does not know if rules are being followed because they never look to find out.

The ***No One Told Me Culture*** comes from a lack of communication channels from workers to management about problems. Problems are dealt with low in the organization and often don't get to the level of leadership necessary to allocate sufficient resources to fix the problems. Thus, management never hears about issues that employees deal with, and often make do with inadequate answers, every day.

The ***Poor Problem Solving Culture*** comes from poor troubleshooting and root cause analysis of problems as well as poor development of effective fixes for problems. Often, only the symptoms of problems are addressed although considerable money and effort may be wasted implementing ineffective fixes. The main symptom observed in this culture is repeat problems.

The ***Backlog Culture*** comes from a chronic shortage of resources and allocation of resources to work other than correction of previous problems. Often this shortage is accompanied by extensive crisis management dealing with issues that should have been fixed but the corrective actions were never accomplished because: "We are too busy around here fighting fires."

The ***Bad Safety Culture*** is a combination of most of the previous cultures. It will definitely include the Git-er-Dun, See-No-Evil, and No One Told Me Cultures and a dose of Poor Problem Solving and Backlog Culture as well. Crisis, blame, and employees with "bad attitudes" will be symptoms of this combination of bad culture issues.

These cultural issues will often be combined with site-specific factors that lead to problems that create crises. The site-specific factors may include:

- Bad procedures
- Poor verbal communications
- Supervisory issues
- Poor training
- Poor human engineering
- No/bad quality control
- Focus on reactive versus proactive maintenance

These issues never seem to go away because ineffective problem solving (Poor Problem Solving Culture) is a part of the culture issue.

ROOT CAUSE ANALYSIS INDICATORS OF CULTURE PROBLEMS

Identifying these culture issues using advanced root cause analysis isn't difficult. For example, using the TapRooT® Root Cause Tree®[7], each shows up as one of the root causes under the Management System Basic Cause Category shown in Figure 1.

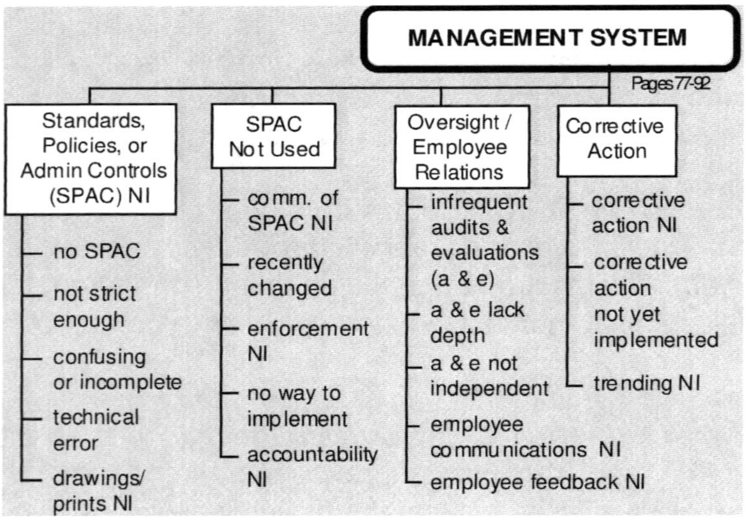

Figure 1: Management System Root Causes
(Copyright © 2008 by System Improvements, Inc. Used by Permission)

For example, the Start-Up Culture issues will be identified with "no SPAC" and "not strict enough" Standards, Policies, or Administrative Controls Needs Improvement (SPAC NI) root causes.

Use of the TapRooT® Root Cause Tree® including the expert system Human Performance Troubleshooting Guide and the Root Cause Dictionary[8], can help lead people investigating incidents and accidents to these culture issues by a systematic process of selection and elimination of potential causes using evidence collected during the investigation.

FINDING CULTURE PROBLEMS: EASY! ... FIXING CULTURE PROBLEMS: HARD.

Finding the root causes of cultural issues is fairly easy. Using advanced root cause analysis it can be performed systematically and repeatably. Why isn't it also easy to fix Safety Culture issues? Quite simply, because they are usually deeply ingrained in the organization and the leadership style of its senior management. They established the Safety Culture. They usually believe that the cause is not the Safety Culture, but rather, bad employees that just need to "be more careful" and "try harder."

To fix these issues, senior management will have to become the sponsor of changing the Safety Culture that they established. They will have to realize that different leadership styles and methods need to be established. And it is the unfortunate job of the root cause analyst (and the sponsor of the improvement program) to show them the impact of the current leadership methods and what needs to be changed.

CONCLUSION

Instead of making culture evaluation a complex, confusing, exercise that can only be performed by people with a culture PhD, one can make the analysis of an organization's culture a systematic process that can be understood by any root cause analyst. The data needed is not a set of complex interviews and surveys. Instead, the data can come from incident/accident data and proactive assessments that the organization already performs.

The difficult part of the analysis is not "understanding the Safety Culture." Rather, the difficult part of the analysis is convincing management to change. Examples of successful Safety Culture change without significant management turnover (usually after a major accident) are rare. But that doesn't mean that the challenge should be avoided. Rather, a careful analysis and use of outside consultants to help senior management realize a new, non-blame oriented vision can make truly dramatic change possible.

REFERENCES

1. Columbia Accident Investigation Board, *Columbia Accident Investigation Board, Report Volume 1*, Government Printing Office, Washington, DC, 2003.

2. HSC Advisory Committee on the Safety of Nuclear Installations, *Third Report of the ACSNI Study Group on Human Factors. Health and Safety Commission,* HSE Books, Sudbury, England, 1993.

3. International Atomic Energy Agency, *International Atomic Energy Agency's (IAEA) Summary Report on the Post-Accident Review Meeting on the Chernobyl Accident*, IAEA, Vienna, Austria, April 28, 1988.

4. US Nuclear Regulatory Commission, "Policy Statement on the Conduct of Nuclear Power Plant Operations," *Federal Register*, January 24, 1989.

5. Institute of Nuclear Power Operations, *Principles for a Strong Nuclear Safety Culture*, INPO, Atlanta, GA, November 2004.

6. Mark Paradies, "How To Identify Culture Issues When Performing a Root Cause Analysis," posted at the *Root Cause Analysis Blog* (http://www.taproot.com/wordpress/2008/06/26/finding-cultural-issues-a-talk-at-the-taproot®-summit/), System Improvements, Inc., Knoxville, TN, June 26, 2008.

7. Mark Paradies and Linda Unger, *TapRooT® - Changing the Way the World Solves Problems*, System Improvements, Inc., Knoxville, TN, 2008.

8. System Improvements, Inc., *Root Cause Tree® Dictionary*, System Improvements, Inc., Knoxville, TN, 2007.

Implementing a Culture of Safety and Operations Excellence at Newport Chemical Agent Disposal Facility

Tulanda Brown, Mark Brothers, Doug Reehl, Pat Guy
Parsons
Newport Chemical Depot
Highway 63 South, PO Box 519
Newport, Indiana 47966

EXECUTIVE SUMMARY

The Newport Chemical Agent Disposal Facility (NECDF) is a one-of-a-kind, state of the art, neutralization facility designed to destroy the Chemical Agent, VX. VX is a rapid acting lethal nerve agent that was produced in the 1960s at the Newport Chemical Depot in Indiana. Parsons was contracted by the Chemical Materials Agency to design, construct, and operate the facility to process 1690 ton containers filled with chemical agent VX stored at the facility. On August 8, 2008, Parsons completed the site mission to safely destroy 100% of the chemical agent stockpile; eliminating a potential terrorist threat and helping our nation achieve international treaty goals.

NECDF experienced a negative safety trend in late 2006 through early 2007. Injuries were not agent related, but ranged from strains and sprains to lacerations. This trend began to reverse in early 2007, as site management began implementation of a Safety Management System that consists of establishing core values and implementing programs such as the Employee Based Safety Program, Job Site Safety Walkdowns, Human Performance Improvement Program, and NECDF Absolutes.

After successful implementation of these initiatives, NECDF went from having the worst Recordable Injury Rate amongst the active Chemical Demilitarization Sites to the best in less than two years. NECDF has also been acknowledged twice by the National Safety Council for achieving one million safe work hours without a lost workday incident for the last two years. Recent statistics indicate that this trend is increasingly positive with no first aid injuries for the month of September 2008.

This paper will outline how elements of the Parsons Safety Management System were implemented at NECDF, describe the events that created the negative safety trend, and describe the initiatives implemented to reduce injuries at the facility.

PROJECT DESCRIPTION AND BACKGROUND

During the 1960s, the Newport Chemical Depot (NECD) in Indiana produced the chemical agent VX (a rapid acting, lethal nerve agent) until a unilateral decree halted production and transportation of all chemical weapons. In 1999, the U.S. Army awarded Parsons the contract to design, build, and operate NECDF, the demilitarization facility that destroyed 4% of the Nation's original stockpile of chemical agent.

As a result of the terrorist attacks on September 11, 2001, the U.S. Army accelerated its schedule to destroy the chemical agent VX being stored at NECD. Parsons was asked to apply its expertise in modularized engineering and construction in order to redesign the facility so it could neutralize the agent VX faster. While the facility was under construction, the process module and materials-handling assemblies were concurrently being fabricated and tested offsite. The process modules were then transported to NECDF and installed in conjunction with final construction – thus saving considerable time.

Full operations of this one-of-a-kind, state of the art, neutralization facility began in May 2005.

NECDF is the first site to destroy VX agent safely using a caustic neutralization technology. Workers first drain the nerve agent from steel storage containers using three special glovebox systems. The nerve agent is pumped out of storage containers into holding tanks, and the agent is then fed into steel neutralization vessels (reactors) where it is neutralized with hot sodium hydroxide and water, forming a caustic wastewater byproduct. The caustic wastewater, also called hydrolysate, is tested at an onsite laboratory to ensure that it contains no detectable agent. The hydrolysate by-product is then transferred into temporary intermodal storage containers, awaiting shipment to a commercial treatment, storage, and disposal facility. The empty steel containers go through a final decontamination step, using another Parsons-designed process. The steel containers are stored onsite until they are transferred to a commercial steel recycling center.

Parsons has recently achieved a significant milestone of destroying 100% of the chemical agent VX stockpile. On August 8, 2008, Parsons and the Army safely rid our nation of a potential terrorist target by removing the risk posed from the continued storage of agent VX in the stockpile.

SAFETY CULTURE PAST

NECDF experienced a negative safety trend having a sum total of eleven different recordable injuries from March 2006 through February 2007. This negative trend began to reverse in 2007, as site management began implementation of a comprehensive Safety Improvement Initiative.

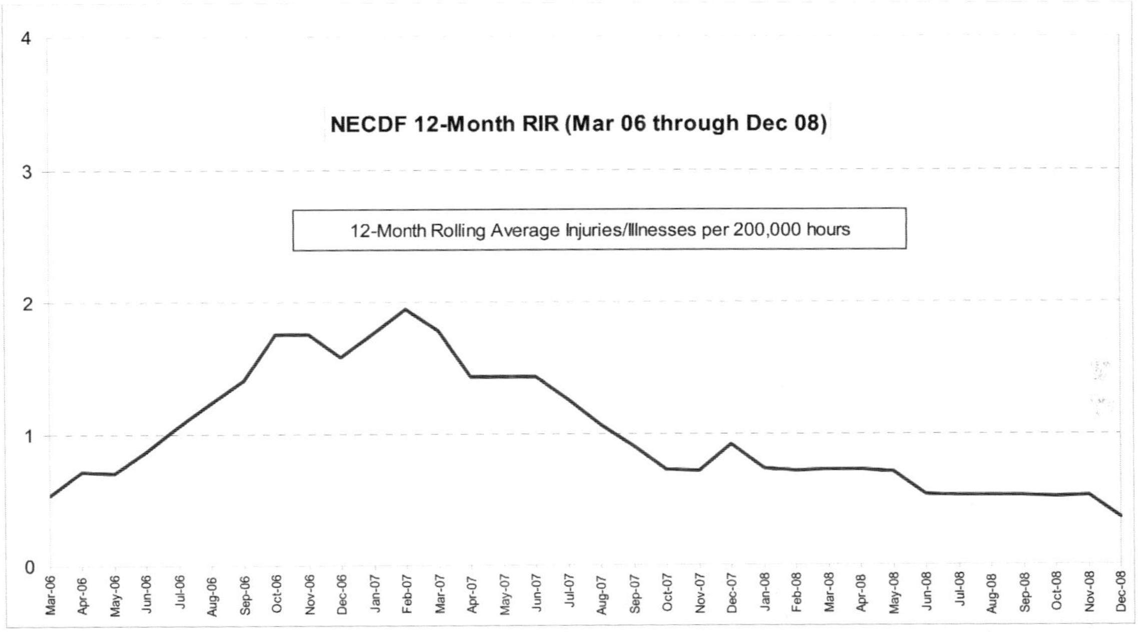

Figure 1 – NECDF 12-Month Recordable Injury Rate (RIR)

NECDF needed to understand why people were not working safely or were taking risks at work and we needed to create a work environment that supports people in working safely. In order to understand the barriers of having a safe working environment, a Stand Down was conducted in February of 2007, and a series of Rock in the Road exercises were held with workforce members. This was initiated with a review of the negative safety results at NECDF since February 2006. Input was solicited without edit. The employees were extremely open and took full advantage of the opportunity to express their concerns. Over 425 individual comments were recorded in these sessions and have been subsequently documented for review. As can be expected with four separate sessions, there was a significant amount of duplication, both within and between the sessions. Barriers to having a safe working environment as identified through a series of Rock in the Roads exercises identified by the workforce were:

1. <u>There was a perception that senior management allowed the pressure for production to over-ride safety concerns.</u>

2. Field personnel did not feel comfortable pausing work for safety issues because of perceived "intimidation" from mid-level and senior managers.

3. The aggressive schedule for chemical agent neutralization created the perception that schedule overrode safety concerns.

4. Maintenance Corrective Action schedules did not promote or support an effective safety culture in that scheduled and planned corrective action work would be substituted for an emergent condition of higher priority and not enough time would be allocated for planning and assessing the risks involved in the "new" work

5. Communication needs to be improved, the most commonly mentioned problem identified within this category was that management changes policy, process, directs deviations from procedures, or changes the plan and does not tell the workforce why, or provide justification for the change, or identify when the change is to be terminated and a return to prior state effective. The issue of management ordered deviation to policy or procedure was a particular bone of contention. Basically, the stated sentiment was; "management tells us we have to follow procedure unless it gets in the way of production and then the direction is to forget the procedure and get the job done".

6. Employee Attitudes and Behaviors was another area that was consistently brought up in each session. Observations included personnel becoming complacent after performing repetitive tasks for four years, personnel wanting to do things "their" way based on previous experience, personnel not following procedure, personnel feeling constrained by procedures, personnel not caring about other employees, and employees not focused or aware of surroundings, etc. Many of these behaviors or attitudes noted were combined with a lack of accountability. In other words, personnel not performing to standard were not held accountable for their actions.

In response to our employees' concerns several programs including; an Employee Based Safety Observation Program, a Human Performance Improvement Initiative Program, a NECDF Absolutes Program, and a Job Site Safety Walk Down Program were initiated and will each be discussed briefly.

NECDF CORE VALUES

The NECDF safety performance turn around was the direct result of establishing core safety values. The various safety programs served to reinforce/support the established core values. In addition, a rigorous response to first aids and near-misses served to decrease the OSHA recordables. Issues and incidents were managed at a lower level and attention was given to fact-finding, causal analysis, and swift corrective action.

Established Core Safety Values:

1. Rigorous implementation of Conduct of Operations Program is not optional.
2. Every employee has responsibility for safety, their safety and the safety of others (establishes ownership and accountability).
3. Training and accountability for employees and supervisors, such as implementation of verify procedures and defined hold points.
4. Recognize and reward safe performance; one initiative was the joint contractor/client determination to expand the employee retention program based on safety performance.
5. Adopt philosophy of "steady as she goes" – find the natural pace for the plant and stick with it, resist temptation to "tweak performance" or experiment with "optimization".

These values were adopted by all NECDF employees including subcontractors. The focus became balancing priorities and ensuring safety was integrated into all aspects of operations.

EMPLOYEE BASED SAFETY PROGRAM

In January of 2007, Parsons' Senior Management at NECDF attended an "Employee Based Safety" overview presentation conducted by corporate and division safety managers from various Parsons' locations. The program was implemented in March of 2007.

Employee Based Safety (EBS) is a peer on peer observation and feedback program. Employees are trained in how to conduct observations and how to give proper feedback in regards to safe and at risk behaviors. EBS is behavior based safety through behavioral science technology. This technology is not new; however, it has been refined in various ways over the past twenty five years and there are multiple versions of this type of program. In brief, EBS uses a set of principles and procedures recognized worldwide for its science- based and effective approach to preventing workplace injuries. The program at NECDF is managed under the following procedures:

- Non supervisory employees will conduct the observations.
- Only volunteer observers are utilized.
- Observers attend a five module training course that lasts eight to ten hours.
- The volunteer observer learns the skill of proper coaching and feedback

- The number of observations conducted by a single observer, and the time involved, is agreed upon during training.
- Permission must be granted from the observed before the observation is conducted.
- Immediate feedback and correction is shared during the observation if "at risk behavior" is observed. Just as important, proactive feedback is shared when safe behaviors are observed.
- The checklist that is used during the observation is filled out by the observer and shared with the observed.
- The information identified on the checklist (safe behaviors/at risk behaviors) is entered into a computer program for trending and graphing.
- The "Golden Rule" while doing observations is "No Name, No Blame."

Once the program was installed, observers received feedback in regards to the quality of their observations. Currently, this is performed on a monthly basis by randomly selecting an observer and reviewing their observation and feedback results with them.

Once per month a meeting is conducted with the entire observer group to ensure we are meeting our goals. Approximately every three months the observer group is reviewed for potential changes in personnel. Observers drop out on occasion, due to lack of interest, etc.

Parsons' observers at NECDF, to date, have conducted over 30,000 observations to date which has contributed to overall injury prevention. The EBS program will remain a mainstay in the safety management program at NECDF during closure activities.

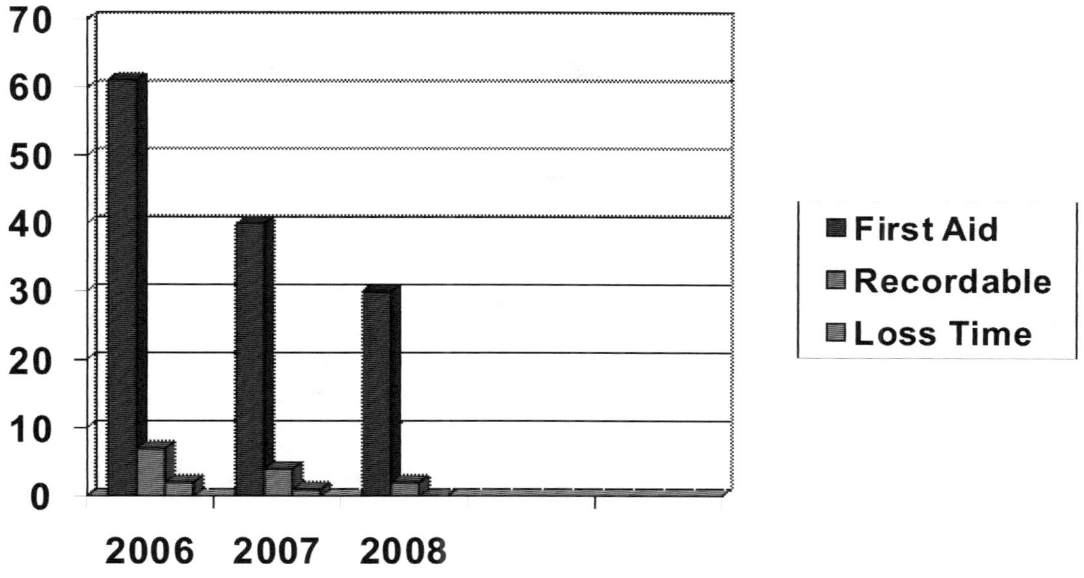

Figure 2 – Newport Injury Results 2006-2008

Since implementation of EBS during first quarter of 2007, we have experienced a 100% reduction in lost time injuries, 78% reduction in additional recordable injuries, and a 50% reduction in first aid injuries.

HUMAN PERFORMANCE IMPROVEMENT (HPI) INITIATIVE

In December 2007, NECDF underwent a number of operational challenges that affected the safe operation of the plant. Parsons paused work and conducted a detailed analysis of the underlying issues causing the incidents. The investigation reports indicated that the root cause of a number of the incidents involved errors by operators and procedure noncompliances by staff. Parsons did not look at these incidents as isolated events but as symptoms to possible underlying organizational problems.

A steering committee was formed to research and implement elements of the Human Performance Improvement (HPI) Program. The committee hired a consultant and expert in HPI Implementation to conduct a formal gap analysis of facility processes and systems and provide a recommended path forward for HPI Implementation at NECDF. The Steering Committee reviewed the gap analysis and recommendations from the consultant and worked with NECDF staff to incorporate the principles into existing programs and training for Operations, Maintenance, and the Laboratory.

There were some examples of HPI already being implemented at Parsons

- Peer review checklists
- Conduct of Operations – Well-defined Conduct of Operations Manual with Limiting Conditions of Operation and flow down requirements
- Technical Inquisitiveness/Questioning Attitude by Operations and Maintenance Staff
- Weekly Operational Emergency Drills with all necessary responders participating.
- Color Coded Procedures – Coded by Type of Procedure
- Command and control by Operations Staff
- Knowledgeable staff familiar with unique processes
- Implement ALARA (As Low as Reasonably Acceptable) for Exposures to Material At Risk
- Integration of Environmental, Safety, and Health upfront
- Pause Work Program
- Formalized Critical Activity Meetings and Pre-Job Briefings

The gap analysis review indicated the potential issues and precursors

- Common Traps of Human Nature – Requirements Backsliding and Summit Fever
- Change Imperative – Control lost in one area could show up negatively in another
- "New" Initiative Syndrome – "Flavor of the Month"
- Concern for arbitrary, inconsistent discipline and punishment

Specific changes were as follows:

1. Revised the Conduct of Operations Course to reflect the principals of HPI.
2. Integrated HPI Principles into the focus training for Operations, Maintenance, and Laboratory Personnel. For example; focus training requested for Operators on the expectation of reader/checker roles could be started with training on the fundamentals of HPI and then using the reader/checker as an example of a tool of Human Performance Improvement.
3. Procured focus training for supervisors to specifically focus on the tools of coaching, communication, situational leadership, and assertiveness. It is important that the supervisor training concepts are supported at all levels of management from top to bottom. For this reason, we required this training for all management positions in Operations, Maintenance, and the Laboratory. The duration of the training was one day for eight hours.

As a result of implementing the programs and emphasizing the principles of HPI, the NECDF did not have another significant incident for the balance of the operational phase.

NECDF ABSOLUTES PROGRAM

In addition to implementing HPI, Parsons also launched the "Absolutes" program at the NECDF. Over the next several months, this program brought attention to the core values we consider absolutely necessary for continuing safe and compliant operation at the NECDF. Parsons highlighted "Absolutes" in the areas of Environmental, Safety, Surety, and Conduct of Operations. These absolutes aligned principles currently outlined in NECDF documentation and were featured in "Toolbox" topics to start Pre Job Briefings each day. The briefings were highly interactive sessions led by operations supervisors, with personnel from all levels encouraged to dialogue during the sessions. NECDF Absolutes posters were posted on site in various locations in the administrative and chemical processing areas. The posters featured photos of NECDF personnel.

The NECDF Absolutes are as follows:

1. SAFETY IS OUR CORE VALUE
    a. Exercise the right to stop/pause work if you believe an unsafe condition exists.
    b. Ensure only certified personnel with demonstrated competence in performing their duties will operate equipment/systems.
    c. Ensure procedural compliance and safe comfortable operations tempo dictate production. The number of evolutions performed concurrently should be limited to allow proper supervisory control.
    d. Ensure all activities on equipment important to safety, affecting operations, changing control indications or alarms, are properly analyzed, documented, and authorized.
    e. Observe LCOs to document plant conditions and to control the plant when safety features are disabled or compromised.
    f. Exhibit technical inquisitiveness and cultivate a questioning attitude.
    g. Constantly examine safety.

2. MAINTAIN CONTROL OF EQUIPMENT AND SYSTEM STATUS
    a. Plant Shift Manager maintains proper facility configuration, including changes to equipment and system status.
    b. Use status boards and logs to identify and track abnormal equipment status or condition.
    c. Install, operate and maintain Temporary Modifications per approved procedures.
    d. Ensure operating personnel have the latest revision to documents necessary for proper control of equipment and systems.
    e. Utilize the Lockout/Tagout system to protect personnel and equipment during maintenance, troubleshooting, and post-maintenance testing.

f. Conduct briefings frequently, for abnormal or emergency operating conditions, to strengthen operator understanding of facility conditions and operating procedures.

3. A PROFESSIONAL ATMOSPHERE MUST BE MAINTAINED so activities performed in the facility remain focused on operations.
   a. Exhibit formal, professional, businesslike behavior to enhance the quality and command and control of operations.
   b. Exhibit high standards of technical and ethical performance to enhance the foundation for safe and reliable operations.
   c. Limit access to operations areas to personnel on official business and potentially distracting activities should be prohibited.
   d. Ancillary duties should not affect monitoring and operation of instrumentation and control panels.

4. IMPROVE PLANT SAFETY AND RELIABILITY by preventing equipment breakdown and maintaining the equipment in a satisfactory condition for normal operation and/or emergency use.
   a. Clearly identify and report unsafe conditions and deficiencies through the Condition Reporting System.
   b. Ensure maintenance activities are identified, initiated, planned, approved, scheduled, coordinated, performed, and reviewed for adequacy and completeness.
   c. Minimize exposure of personnel to hazardous materials and conditions.
   d. Ensure all maintenance performed by qualified and trained personnel.
   e. Ensure individuals performing work or responsible for work are adequately trained and have a clear understanding of work scope and the effect of the work being performed.
   f. Never By-pass or deactivate safety controls and interlocks.
   g. Perform maintenance only in accordance with detailed, approved procedures and permits.

5. COMMUNICTATE IN A FORMAL AND DISCIPLINED MANNER.
   a. Use Three-way communication to ensure accuracy of all verbal direction. Establish communication; Transmit message; Repeat back message; Confirm repeat back.
   b. Ensure communications are to be accurate and complete; most importantly, understood.
   c. Ensure communication systems are properly used. Adhere to business-only use of in-plant communications equipment.

6. MAINTAIN A STRONG AND PROGRESSIVE SURETY CULTURE
    a. Handling, neutralization, and accountability of chemical agent is strictly performed in accordance with applicable policies, plans, and procedures.
    b. Aggressively pay attention to detail and promptly report Chemical Personal Reliability Program certification issues to Government certifying officials.
    c. Maintain physical security measures to preclude unauthorized access to Limited and Exclusion areas or access to, or use of, chemical agents.
    d. Maintain an effective emergency response capability and continuously evaluate its effectiveness.

7. PROTECT OURSELVES, EACH OTHER AND OUR ENVIRONMENT.
    a. Know how your work activities impact the environment and incorporate pollution prevention and waste minimization into these activities.
    b. Maintain positive control of agent and hazardous materials at all times.
    c. Minimize waste generation.

JOB SITE SAFETY WALKDOWN (JSSW)

Job Site Safety Walkdowns (JSSWs) are conducted to ensure that employees walk down the work, in the area that it is being performed, just prior to work commencement. Employees are encouraged to always consider the environment (e.g., night, day, dry, wet, icy, etc.) and determine the tools and equipment required to safely perform the work. JSSWs are not only a safety tool, but also serve as an excellent work planning tool. JSSW is designed to be completed by personnel who are physically performing the work after walking down the work site and reviewing the steps needed to perform the task.

Supervisors were given specific instructions to ensure the JSSW process was effective and consistently applied throughout the site. Instructions are outlined below:

JSSW process consists of three steps:
(1) Identify Work Steps,
(2) List Potential Hazards, and
(3) List Steps to Eliminate Hazards

In defining the job it is important to:

(1) BE SPECIFIC: Don't try and analyze the whole project. Break it down into job/tasks. Look at one job/task at a time.
(2) IDENTIFY THE TEAM: Every individual on the job/task is part of the JSSW. If you can't get the team together to do a JSSW, your scope is too wide.
(3) BE TIMELY: The most important task to analyze is the one currently being performed. A job spread over several days or crews requires more JSSWs.

As the steps are outlined remember:
(1) REVIEW THE JOB: Involve the whole team. Review procedures, plans, work orders, and OPFIs.

(2) BE LOGICAL: Identify logical steps. Don't get buried in details. More than 15 steps may mean too wide a scope or to much detail.

(3) BREAK IT DOWN: List the steps in order you plan to do them. Number each step. Be descriptive in steps. Start each step with an action word.

In identifying the hazards:
(1) GET REAL: Call on Team Members' experience. Be specific on hazards and avoid catch alls like: falling hazard, agent hazard, etc. Avoid philosophical debates that go nowhere.

(2) BE METHODICAL: Walk through each step. Address one hazard at a time. Number each hazard using sub letters (e.g., 2a, 2b, etc.).
(3) BE THOROUGH: Use realistic scenarios. Identify existing hazards – to include environmental hazards.

(4) ADDRESS ALL HAZARDS: Work identified hazards one at a time. Number each hazard control measure to be employed. Use sub-numbers if needed (e.g., 2a1, 2a2, etc.).

(5) BE PRACTICAL: Call on Team Members experience. Try to specify control measures that are doable with available resources. Call on experts when in doubt about controls.

(6) BE SPECIFIC: Avoid catch all controls like: Be careful, wear proper PPE, stay alert, etc. Assign required actions by NAME (e.g., Fred will guard the floor opening).

NECDF currently has processes to document risk mitigation through either Job Hazards Analysis (JHA) for implementing operating procedures or Job Safety Analysis (JSA) for maintenance activities. Both analyses are formal and are developed in a team environment to identify hazards associated with work tasks and outline the planned strategy to mitigate those hazards. The Job Site Safety Walkdown is a separate process and is very important in that it allows workers who are going to be involved in the work task to take one last look prior to starting work. We believe this has had a positive impact on preventing injuries at NECDF.

SAFETY MANAGEMENT STRATEGY AND PROGRAMS FOR SITE CLOSURE

The strategy for safety management performance during closure was developed during May and June of 2008. After a complete audit of existing programs, a review of all reported loss events, employee and management interviews, and finally an offsite closed door session with managers, supervisors, and representatives of the safety department, the safety program development and strategy for closure began.

The strategy employs a number of programs that were implemented during operations but also adds enhancements to incident investigations, management led safety initiatives, and employee involvement.

Incident Investigation - Event Cause Analysis or "Fact Finding"

The first addition to the closure safety program came by way of a modification to accident/incident investigation procedures that had been in place since inception. After a review of the procedures there was confusion in the way supervisors and managers identified the causes of loss events. The first rule in post event prevention is to have a comprehensive, yet simple, procedure to help management identify the causes of any loss event.

First, identify the "symptoms" or the "immediate" causes of the event. Supervisors and managers were taught, using a provided checklist, to review examples under the following:

- Substandard Acts / Behaviors and Substandard Conditions
  Once examples are identified simply ask why? Why did the substandard act, behavior, and /or condition take place or exist. This takes the supervisor or manager to our next provided checklist entitled "basic causes." At this time they review examples under the following:

- Personal Factors / Job Factors
  Once basic causes are identified, the supervisor or manager has a "lack of control" list to review to further help them identify where potential changes may need to be made in any current program or procedure. If at risk behavior was a key cause, it will be identified in this process as well.

In addition, using this system paves the way for informal or formal root cause analyses if necessary. In fact, root causes can and are identified using this system, eliminating the need to use a more formal system ninety percent of the time.

To add to this program, we took this system and helped management establish an open forum during cause analysis. Instead of a private one-on-one cause investigation, management invites all involved in the loss event to openly participate in finding the causes using fact finding, not fault finding, thus, creating the name of the process "Fact Finding."

This forum is open and creates a proactive environment in order to find the facts, or the actual causes to any loss event or near–miss. Constructive remedial actions result from effective fact finding sessions. Now, by identifying the true causes of any type of loss event, management is learning to prevent further loss. This is formally called "Post Contact Loss Control." This is a new concept for Newport that involves all levels of management and hourly employees.

Management Led Safety Initiatives

Supervisors and managers at Newport were given specific safety management responsibilities for Closure. This shifts responsibilities traditionally held by employees to senior and front line level supervisors, where it belongs. Safety professionals continue to conduct inspections, assist in safety meetings, and participate in all other aspects of the program, however; with senior, middle and front line management involvement, there is more safety program ownership. The role of the safety professional here is managing the manager.

Meetings were held with all crews and supervisors to introduce each new initiative. Each new program or enhancement to an existing program was explained in detail. In addition, we outlined the methods being used to measure the effectiveness of the programs.

Management led programs are outlined below:

<u>All Hands Safety Meetings</u>
Once per month the VP and Deputy Project Manager for Closure meets with all DDD crews and present a safety topic and safety performance update. Hourly employees receive safety information directly from a senior manager. This shows support and creates trust. The safety professional develops the agenda, ensures the senior manager understands the presentation, and assists Senior Leadership during the presentation.

<u>Leading Indicator Audits</u>
One per month the VP and Deputy Project Manager for Closure conducts leading indicator audit of the project using a provided checklist. For so many years we have

concentrated on lagging indicators; i.e., how many recordable events etc. Using leading indicators is more proactive. NECDF now tracks the number of inspections conducted, the number of safety contacts, and number of employee observations. Measuring the leading indicator and advertising it, plus, auditing the leading indicator program from a senior manager perspective shows support and leadership from Senior Management.

Senior Manager Quarterly Safety Inspections
Once per quarter, selected members of senior management walks a segment of the project with a safety professional during formal quarterly inspections. This is a tool to get the senior managers active in conducting inspections and observing behaviors and conditions that could cause a loss. More support, more buy in, better program.

Supervisor Weekly Safety Inspections
Supervisors inspect the facility using a provided checklist on a weekly basis. This is a tool to utilize the front line supervisor to observe behaviors and conditions that could cause a loss.

PA Safety Announcements
On a daily basis supervisors use the intercom system to announce a safety message to the entire project. The supervisor participates in the daily activities of communicating safety.

Accident Prevention Council
The VP and Site Project Manager meets monthly with employees representing all departments in regards to overall safety management. This program was retained for closure. Senior Management involvement in resolving employee safety concerns is the goal.

As mentioned earlier, survey results indicated employee involvement in the safety process is necessary to ensure personnel are constantly aware of potential hazards in their work area, personnel have a sense of accountability and ownership for their safety and the safety of others, and that they do not become complacent in performing work tasks.

Safety programs for hourly employees

Safety Contact Program for Hourly Employees
Employees are asked to make a formal safety contact with an employee of their choice and document the name. A safety contact may consist of an employee asking a coworker, "Do you have all required PPE to safely perform your duties of the day?"

This is an extra peer check to ensure the conditions and behaviors for a safe work environment exist.

Pause Program
Employees will pause if and when they feel a substandard condition exists that will affect their safety. A formal procedure is in place. This program is retained for closure; however, the emphasis that was placed on the importance of pausing during closure is unsurpassed.

Employee Based Safety
Peer on peer observations performed daily by hourly employees only. As indicated earlier, this program was very successful for operations and will continue during closure.

Safety Committees
Employee safety committees were formed during operations and are being continued throughout the closure phase of the project. Each committee meets periodically to review safety initiatives and discuss safety concerns in their area.

Pre Job Walk Down/ Dry Run
Employees participate in preoperational demonstrations to ensure safe execution of work prior to commencing the actual activity. Job Hazard Analysis, Job Safety Analysis, and Job Site Safety Walkdowns are used to document the safety analysis of work tasks.

Safety Banners
Employees submitted a safety slogan using zero injuries as a guideline. A winner was chosen and the banner is currently displayed with the winning slogan for Closure, "Wrapping it up with Safety". Safety stickers and posters were also given to supervisors and employees advertizing the goal of zero incidents and the closure slogan.

Charlie Morecraft Presentation

To kickoff our safety awareness campaign for closure Charlie Morecraft was invited to speak to nearly 600 NECDF employees at the ISU Civic Center in September 2008. All of Parsons corporate leaders attended the presentation; this included our Chief Executive Officer, Chief Operations Officer, Corporate Vice President for Safety, and Global Business Unit President. Management and employees from the Chemical Materials Agency, our client, also participated.

Charlie's story is true. He received third degree burns over 50% of his body during an event that occurred at an Exxon refinery in 1980. His story explains how he caused the event (fire), his negative attitude towards safety, his five years in numerous hospitals and subsequent surgeries, his father's death shortly after the event, and so on. His message is about personnel accountability and responsibility regarding safety on and off the job.

Charlie's message was powerful and well received by everyone in attendance. As we entered the Closure phase, safety awareness and individual accountability has helped prevent injuries. Charlie's message also had personal effects on some of our employees. Many employees indicated that his story has changed their lives.

VOLUNTARY PROTECTION PROGRAM IMPLEMENTATION

The success of the programs described above drove us to consider obtaining VPP Star Status through Occupational Safety and Health Administration (OSHA). We have submitted our application and expect the formal review to be conducted in the spring. NECDF is currently implementing all four elements of OSHA's Voluntary Protection Program. Our implementation strategy is outlined below:

Element #1: Management Leadership and Employee Involvement - Parsons must show top-level management commitment to occupational health and safety. The management system must integrate the authority and responsibility for employee health and safety into the management system of the organization and must involve employees in the process.

NECDF Implementation: Safety and Health Policy, Program Managers Accident Prevention Council, Safety and Health Committees, Employee Based Safety Program, Condition Reporting System, Individual Safety and Health Performance Objectives.

NECDF employees at all levels must be involved in the structure and operation of the safety and health program and in decisions that affect employee health and safety. Employee participation is each employee's individual right and responsibility to notify appropriate managers of hazardous conditions and practices.

NECDF Implementation: Job Hazard Analysis (JHA, JSA, & JSSW), Employee review of procedures, Employee Safety Committees, Safety Daily Reports, and Accident Investigations.

Element #2: Work Site Analysis – Management of health and safety programs must begin with a thorough understanding of all potentially hazardous situations and the ability to recognize and correct hazards as they arise.

NECDF Implementation: Job Hazard Analysis (JHA, JSA, & JSSW), IH sampling/monitoring, Pre-existing and baseline conditions assessments, New/Modified equipment process reviews.

Element #3: Hazard Prevention and Control - Hazards and potential hazards identified in the worksite analysis must be prevented or minimized by the implementation of effective controls and measures.

NECDF Implementation: Safety Design Requirements, HAZOP, Engineered Controls (HVAC, process designs, etc.), OSHA Operational Safety Controls, Limiting Conditions of Operation, Administrative Controls (Procedures, etc.).

Element #4: Safety and Health Training - Managers, supervisors, and employees must know and understand the policies, rules, and procedures established to prevent exposure to hazards. Training for health and safety must ensure that responsibilities are understood:

NECDF Implementation Examples: Parsons safety training (START and SHARP), Site Orientation Training, Command and Control Training, Employee Based Safety Training, Job/Facility Training (i.e. fall protection training, fork truck operations, respirator training) Safety Meetings, and Accident/Incident reviews.

BENEFITS OF NECDF SAFETY PROGRAM INITIATIVES

Many benefits were realized as a result of the implementation of the programs described. Not only did we have a significant reduction in injuries, employee morale was improved and the safety culture was adopted by entire site, including subcontractors. NECDF has been recognized by several notable organizations for exceptional safety performance. We truly believe that in order to sustain our excellence in safety NECDF obtained management buy-in, built trust amongst employees at all levels, and developed an effective system of self assessment and accountability.

From Training to Human Reliability: Lessons Learned during the Implementation of a Human Reliability Function in the Pharmaceutical Industry.

Ginette M. Collazo, Independent Consultant, San Juan PR, ginette.collazo@gmail.com, 787-530-0200.

Pharmaceutical manufacturing sites compete constantly, even within the same company in order to maintain production volumes and avoid closures. To attain a status of a *high performing organization* manufacturing sites are implementing strategies and practices like operational excellence and performance reliability. In order to achieve this goal, innovative approaches to traditional processes need to take place. This paper discusses how by redesigning the training function and by focusing on the expected result (human reliability) instead of the how (training), a reduction of more that 60% of human error events in pharmaceutical manufacturing facilities was achieved.

THE PROBLEM: REDUCE HUMAN ERRORS

The ultimate objective of a Training function is to assure that people do things right, people do the right thing, and people do it in an efficient way. Training, being the process intended for transferring knowledge, skills, and abilities certainly plays a very important role in the way people perform. Nevertheless, people's behaviors are influenced by a lot more. Constantly we are faced with the reality that people, even though they know what to do to either lose weight, lower their cholesterol, or be healthy just don't do it. Everyday people engage in risky behaviors, even though they are knowledgeable of the possible consequences. Knowledge by itself does not modify behavior.

JUSTIFICATON

Organizations want people to be reliable when they are executing day to day operations. Human reliability defines the ultimate goal of any training event. By changing the focus to reliability instead of training, efforts are directed towards the right actions. A human reliability approach focuses on all of those elements that could impact behavior besides the individual himself, like the clarity of instructions, sequence of actions, clear communication, resources availability, layout designs, and work tools, among others. To ensure flawless execution, organizations could broaden the scope of the training groups from a traditional training function focusing on learning strategies exclusively as a mean to improve performance, to one that addresses other areas that impact behavior as well.

METHODOLOGY

To ensure the correct actions are implemented a diagnosis was performed. This diagnosis consisted in the analysis of failure investigations. A random sample was selected for a 12-month period. These investigations were evaluated considering the following criteria: correct problem defined, correct root causes/causal factors identified, and appropriate immediate, corrective and preventive actions. This step consisted of challenging and validating each of the five elements considered as acceptance criteria. After performing this evaluation, corrections were made to the investigations and events where the criteria were not met.

Once all the correct causal factors/root causes were identified for all events, these were categorized and coded in four iteration levels (as a deductive process from a general to specific view). The first level identified the causal factor type, the second identified the root cause category, the third level identified the near root cause, and the last level the final root cause(s).

Data from six manufacturing processes was considered in the data analysis.

RESULTS

1. More than 80% of failures were related to errors.
2. 8-10% of error events were related to training (knowledge, skills, and abilities).
3. More than 50% were related to procedures, mostly incomplete procedures or confusing procedures.
4. Less than 10% were related to "the individuals" executing the tasks.
5. Approximately 30% of events were distributed in other categories like equipment failures, poor administrative management systems, supervision, and communications.

Findings in the investigation process

1. Problem identifying the event/failure.

2. Identifying one root cause, but failing to identify multiple factors that contribute or cause the failure.

3. Confusion determining the number of iterations (depth) of the root cause/causes determination process.

Findings in the Correction and Prevention Approach

1. Immediate actions were used in place of CA's (corrective actions).

2. CA's (corrective actions) were used in place of PA's (preventive actions).

3. There were no real preventive actions.

4. Most events used training as a corrective or preventive action for human performance deviations.

5. Trending and tracking mechanisms were non-existent.

6. There were no prediction actions/initiatives.

DISCUSSION

Human error constitutes a considerable number of failures in manufacturing facilities in the pharmaceutical industry, impacting efficiencies. Training is very effective for preventing errors when there are new employees, new processes and/or changes are made to existing processes. Training has not shown to be effective on failures that are not associated with either a lack of knowledge, skills, and abilities or when the level of awareness needs to be increased on critical tasks. Nevertheless, the training method and contents needs to be customized depending on the intention. For skills and abilities, on the job or practical training is appropriate. To increase the level of awareness, people need to understand the consequences of failure and science or reasons for specific requirements of a task. On the other hand, the data indicates that training is responsible for a limited number of deviations or failures. This evidences the fact that, since pharmaceutical manufacturing sites mostly use training to correct or prevent these failures, lots of resources and effort are being placed in a very limited part of the problems, affecting not only the efficiencies, processes, costs and the individuals involved, but also the reputation of training professionals that are not being "effective".

Also, most events are identified as procedure related, either because even though there is a lot of information included in these documents, the true important information is missing. This can also be considered as a systemic problem, which in turn would be categorized under management systems related to document formats and risk/task reviews.

Another element to be considered is the fact that less than 10% of the events were related to the individual involved. This is particularly important because most organizations consider a human error to be the individual's fault. This piece of information returns the responsibility of avoiding mistakes to the organization. By designing/ re-designing systems and conditions, failures can be prevented.

Organizational Actions Taken

- Created an interdisciplinary group that had the necessary knowledge to address human-process interface (hardware, equipment, components, systems and thinking process). These professionals included engineers, psychologists, scientists, teachers and other technical experts.

- Provided training on human behavior, instructions development, training and investigation processes:
  - Psychology and physiology of human behavior.
  - Human Error prevention techniques.
  - Risk Assessment
  - Procedures design and technical writing.
  - Training design and delivery, and training effectiveness evaluation techniques.
  - Investigation and root causes determination process.
  - Organizational design.
  - Statistics

- Incorporated Human Reliability experts into the deviation investigation processes.

- Developed a human error investigation process in which Human Error was treated as an event. Not the cause of failures.

- Created a corrective action and preventive action (CAPA) effectiveness program.

- Created a metric mechanism for trending and tracking human performance deviations.

- Created a cultural change/integration program to assure institutionalization.

Specific Actions Taken

- Redesigned training courses to ensure training programs capture critical areas.

- Revised critical procedures to ensure well-written work instructions that take into consideration the way people think and the mental processes that take place when decoding messages.

- Identified critical tasks and steps to ensure training programs were designed highlighting the criticality and the consequences of each task/step.

- Designed training that prepared people for managing unexpected situations.

- Developed programs that considered elements like level of attention, motivation and situation awareness needed for some tasks.

RESULTS

The 60% reduction on human error events was calculated based on the baseline rate (.047 or 4.7%) before the implementation of the above mentioned actions. The baseline rate was calculated by dividing the number of human related events (88) by lots manufactured (manufacturing) and tests performed (laboratories) (1869).

$$\frac{\text{\# Human Error Events}=88}{\text{\# Lots manufactured/tests}=1869} \quad .047 \text{ X } 100 = 4.7\%$$

Using the same formula, the metrics were tracked each month to see improvements, or if necessary, make changes to the original strategy. These metrics were kept for all critical stages in the process (manufacturing stages and testing). Since events were being categorized by root causes and causal factors identified per events, systematic issues were identified and addressed first and isolated instances were addressed later.

The systematic issues identified were related to procedure development. By redesigning the process to develop procedures, i.e., task analysis and what if analysis, procedures were re-written assuring all critical steps had specific instructions. Also, the training program was redesigned to include modules related to the science behind the manufacturing process, making sure employees had the information needed, when judgment was required and unexpected situations arose. An on the job qualification program was established for critical jobs in which mastery of performance was required before employees were released for independent work. The last systematic change was related to verifications required in some "critical tasks". Many steps required verifications, decreasing the level of awareness when true critical tasks needed to be verified for error recovery. These were reassessed and only those tasks that truly needed to be double checked continued to require verification. At the end of the year, the human error rate was .019 or 1.9%, a total of 42 events related to human error in 2186 lots manufactured.

$$\frac{\text{\# Human Error Events=42}}{\text{\# Lots manufactured/tests=2186}} \; .019 \text{ X } 100 = 1.9\%$$

Therefore, a reduction of approximately 60% in human error events was achieved.

Improving Operating Discipline Through the Successful Implementation of a Behavior Based Safety (BBS) Program

Bruce K. Vaughen
Cabot Corporation
700 E US Highway 36
Tuscola, IL 61953

Kelly J. Lock
Cabot Corporation
700 E US Highway 36
Tuscola, IL 61953

Timothy K. Floyd
Cabot Corporation
700 E US Highway 36
Tuscola, IL 61953

ABSTRACT

This paper describes the improvement of a manufacturing site's safety performance that can be attributed, in part, to the successful implementation of a behavior-based safety (BBS) program. As has been discussed in prior papers, improving operating discipline (improving the conduct of operations) will contribute to reducing process safety, health and environmental risks. The BBS program implementation, with its expectation that each employee complete and submit at least one safety observation a month, coincided with several other significant safety initiatives. These other initiatives required employee involvement from all levels of the organization, including implementing Process Safety Management (PSM) elements, implementing the site's "Principles of Operation," and performing Job Safety Analyses (JSAs). This paper will describe how the site's BBS observation program has evolved over the last decade, how several "human element" issues were overcome, and provides specific examples of some of the changes implemented as a result of the observations. In conclusion, this paper shows how increased personnel involvement has improved safety awareness, knowledge, and commitment to help reduce manufacturing risks.

"You can observe a lot just by watching." - Yogi Berra

1- INTRODUCTION

This paper will describe the current Cabot Behavior Based Safety (BBS) Observation program at the Tuscola, Illinois, Fumed Metal Oxide (FMO) manufacturing site, provide a brief summary of the site's other safety programs, and conclude with metrics that show improvements in the site's safety performance. Success of this program has hinged on three basic tenets:

1) Everyone on site is involved and responsible;
2) The observations are kept anonymous; and
3) There are no negative repercussions.

A consistent approach and upgrades to Tuscola's BBS Observation program has ensured its sustainability and its success over the years.

The overall goal of all safety efforts is to reduce the risks associated with both the process and the health-related hazards. One useful way of describing the safety-related methods to reduce these risks is to define the risk with the following equation:

$$\text{Risk} = \frac{(\text{Frequency}) \times (\text{Consequence})}{(\text{Conduct of Operations})}$$

In this equation, the risk is directly related to both the frequency of a possible hazardous event and the event's potential consequence. The frequency is often determined by the effectiveness of engineering and administrative controls, such as safety systems and operating procedures. Both improved engineering and administrative controls will reduce the frequency and hence, reduce the risk. The consequences are characterized by the hazardous properties of the materials (e.g., flammable, toxic, etc.) and the hazards of the process (fires, explosions, or toxic releases). Reducing the consequence, such as those defined within inherently safer design principles, will reduce the risk, as well.

As is defined in this risk model, the safety risk is inversely related to the term called the "conduct of operations." This term reflects the "human element" within the risk equation. As examples, safety systems and operating procedures which are designed to reduce the risk (as is noted in the numerator) must be sustained, maintained and followed to ensure 100% effectiveness. Hence, the overall risk will increase if systems degrade or short cuts prevail such that the engineering or administrative safeguards may fail: the "conduct of operations" term is less than its potential level, the denominator in the risk equation will be lower and the risk increased.

In addition, the "Safety Pyramid," shown in Figure 1, helps visualize how safety efforts can be combined to help reduce injuries. This pyramid contains different levels, beginning with a representation of the number of "foundation" level events (At-Risk Behaviors, the largest "area" and hence, the largest number of events) through to the top level, a representation of the number of fatalities (the smallest "area" and hopefully, the least number of events). It is one of the goals of a behavior-based safety program to attack the base of the pyramid: to reduce the number of at-risk behaviors.

Figure 1. The Safety Pyramid – Industry Statistics

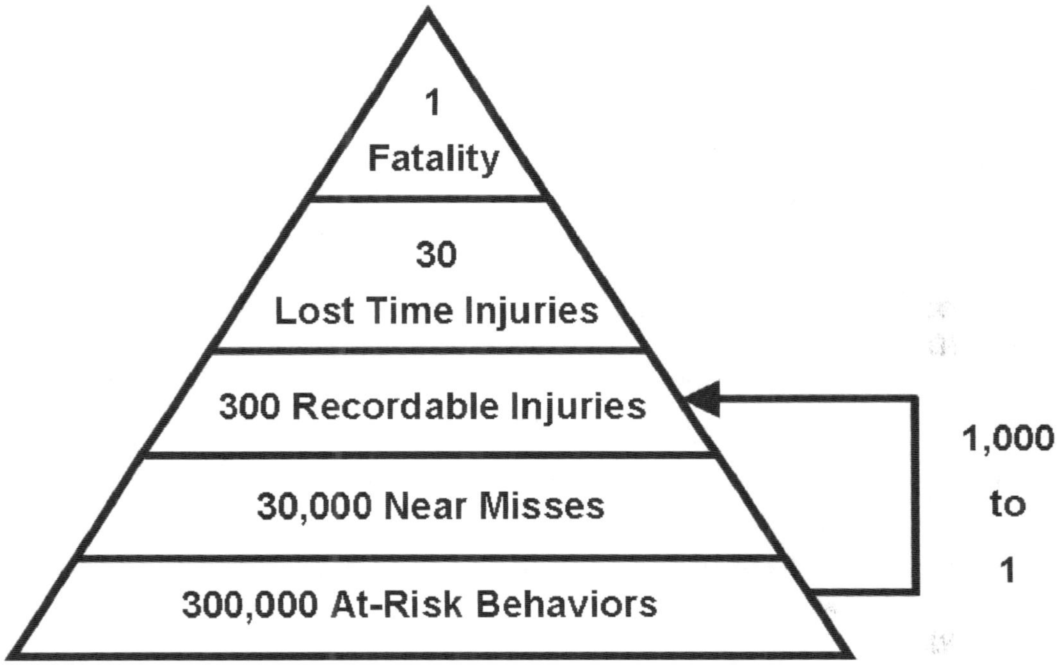

2 - THE TUSCOLA BEHAVIOR-BASED SAFETY (BBS) PROGRAM

The goals of the Tuscola BBS Observation program are to ensure that all employees are aware of the importance of safety, have at least one monthly conversation with one another an some aspect of safety behavior, and that they address any at-risk behaviors appropriately in a timely manner. The expectations of the program include active participation by all employees with one written quality observation submitted per month per employee. Long term benefits of the program include eliminating at-risk behaviors (reducing risk), improving communication skills and a sustaining awareness of safe work practices. The program is designed to help each employee take care of themselves and their co-workers.

This section of the paper describes the current aspects of the Tuscola BBS Observation program that have been developed to meet these goals. It describes the BBS observation program, how the observation information is tracked, how at-risk observations are evaluated and responded to and then concludes with how the BBS observations and results are shared with the site. The second section of this paper briefly describes other risk-reduction approaches to help provide a historical perspective of the other initiatives that have contributed in one way or another to improve the site's overall safety performance.

2.1 - The Current BBS Observation Program

Tuscola initially chartered the BBS Steering Team to direct a voluntary employee observation program and trained the site employees with this premise in 2000. The "S.A.F.E." program was designed to highlight "Safety Awareness For Everyone" and banners and literature distributed across the site contained the colorful logo shown in Figure 2. However, in 2004, the Facility General Manager (FGM) mandated that everyone perform and document at least one observation every month after an increase in severe incidents during 2002 and 2003. Several key learnings by the BBS Steering Team since then have been incorporated into the current training package and are noted below.

Figure 2. The Tuscola Safety Awareness For Everyone (S.A.F.E.) Logo

2.2 – The BBS Steering Team

The BBS Steering Team meets monthly to discuss the observations from the prior month. The BBS Steering Team has at least one member from every department - operations, maintenance, technical, laboratory, logistics and information systems – which provides great inter-departmental interaction. One key discussion item during these meetings is ensuring that any at-risk behaviors noted in the observations are understood (well documented) and have been adequately addressed. In addition to spending time to perform their own observations, these members also volunteer as reviewers of their peers' submitted observations to ensure the quality of the observation. The BBS Team members also attend behavioral-based training sessions and off-site conferences to learn from other companies, and provide BBS training and coaching on site on an as-needed basis.

2.3 - A Summary of the Observation Training Program

The current training package is designed to orient and train new site employees on the BBS Observation program's goals. New site employees receive this orientation within 30 days on site, and are expected to begin documenting their observations monthly once trained. Each employee performs a "dry run" by making an observation and filling out the observation form with the trainer during their initial training. The current training package has captured and shares the program's learnings, having evolved from the initial site communications package in 2000, the site's upgraded and mandatory program implementation package in 2004, and BBS Steering Team program upgrades and refreshers from their safety meeting communications. The outline summarizing the material covered in the BBS Observation training package is shown in Table 1. Each employee is provided with a special BBS Observation note pad and laminated cards during their training (the card is shown in Figure 3).

Table 1. Summary of Material Covered in the BBS Observation Training Package

1	Welcome and Purpose	
2	Observation Definitions	
3	The Observation Procedure	
		The Observation Card and Notebook
		Observation Attitudes
		Observation "Fouls"
		Observing Behaviors Versus Conditions
		Responding to At-Risk Observations
		Writing Quality Observations
4	Observation Practice Run - Field Tour	
5	Observation Results	
		Reducing Risk
		Discussions on Safe Behavior

Figure 3. The BBS Observation Procedure Card

S.A.F.E.

CABOT SAFETY

SAFETY AWARENESS FOR EVERYONE

EMO TUSCOLA

CABOT CORPORATION

Behavioral Based Safety
Observation Program

OBSERVATION PROCESS

1. Announce presence
2. Ask permission
3. Explain process
4. Observe behavior
5. Check boxes on form
6. Write explanation
7. Review observation
8. Give feedback
9. Thank co-worker
10. Turn in

QUALITY OBSERVATION

1. Describe behavior, include a brief description
2. Describe behavior, do not include conditions
3. Details – what, when, where and why of behavior
4. Describe feedback, what was discussed ("We discussed, "I spoke to")
5. Describe changes of at-risk behavior, if needed

The steps to the observation procedure are listed in Figure 3. Note that the observer must announce his/her presence first and ask for permission before beginning the observation (step 4). This prevents the "gotcha" observation, where an employee is caught with an at-risk behavior and is written-up as a result. As is strongly noted during the training, and ensured through the "quality" review before being accepted by the BBS Team, the observation program is a positive attitude program. The BBS Team will not accept and document any "foul" observations (see Table 1). The credibility of the program relies on open discussions between each employee during the observation. Then, after the discussion and approaches to reduce or eliminate the at-risk behavior, the observation is documented (steps 6 and 7). After writing the observation's discussions, the observer concludes with thanking the person who was observed for their time and discussion.

2.4 - The Observation Form

The Cabot Tuscola site BBS Observation form is shown in Figure 4. This one-page form has changed since its introduction, having evolved into the site-specific form in use today. Improvements to this form have been based on BBS program-specific surveys, individual employee feedback, and BBS Steering Team reviews and upgrades. The form contains several areas that must be filled out, and observers that complete each area correctly have completed a quality observation. Each of the areas of this form is described in more detail below.

A submitted form is considered a quality observation when there is a clear and detailed description that documents the behavior at the time of observation (including the 4 W's - What, When, Where, Why), that documents a clear description of the discussion between the observer and the person being observed about the observation, and that clearly documents the action or actions taken to correct any observed At-risk behaviors. Quality conversation examples include statements beginning with: "We discussed...", "I spoke to...", and "I reviewed the observation with...." On the other hand, an unacceptable comment would be: "I thanked the employee for their time, and they thanked me in return."

As is shown in Figure 4, the first section documents the observer, the month and year, the week day, which department the person being observed belongs to, the shift, and the weather conditions. The second section is termed the "Critical Behavior Index[1] (CBI™)," where the observer checks boxes pertaining to the activity being observed. The boxes, a visual checklist, cover these three areas for either "Safe" or "At-Risk" behavior: Personal Protective Equipment (PPE), Procedures and Body. Definitions for these fields are provided during the training. The next section documents the discussion of the observation with the person who was observed. It details any of the

Figure 4. The Tuscola Behavior Based Safety (BBS) Observation Form

Observer:_________________________ **Date of Observation:**________________
(month) (year)

Day: (Circle One) SU MO TU WE TH FR SA

Who Is Being Observed (Dept): _______ **1.** Production, Electronics (A/B/C/D/E/PP) **2.** Quality / Technical **3.** Maintenance **4.** Office (Administration / Division) **5.** WPS **6.** Production, Niche (TDC/Treated/Raw Materials) **7.** Contractors

Shift: (circle one) Day Night **Weather Conditions:** (circle one) N/A Calm Windy Hot Cold Rain Snow / Ice

1)	BODY	SAFE	AT RISK
1.1	Walking		
1.2	Eyes on Path		
1.3	Body Posture		
1.4	Line of Fire		
1.5	Ergonomics		
1.6	Lifting		
1.7	Eyes on Work		
1.8	Pinch Points		

2)	PROCEDURE	SAFE	AT RISK
2.1	Tools & Equipment Use		
2.2	Ladder Usage		
2.3	Work Instruction		
2.4	Permits / LOTO / JSA		
2.5	Communication		

2)	PROCEDURE (CONT.)	SAFE	AT-RISK
2.6	Pre & Post Job Inspections		
2.7	Barricades & Warnings		
2.8	Chemicals		
2.9	Housekeeping		
2.10	Vehicle Use		

3)	PERSONAL PROTECTIVE EQUIPMENT (PPE)	SAFE	AT RISK
3.1	Respiratory Protection		
3.2	Gloves		
3.3	Eye and Face		
3.4	Hard Hat		
3.5	Hearing Protection		
3.6	Fall Protection		
3.7	Long Sleeves		
3.8	Foot Protection		
3.9	Tyvek Suit		

What is the task and *Behavior* being observed?

Describe the location:

Are there any *At-Risk Behavior(s)*? Describe:

What did you and the person being observed discuss during the review of the observation?

How was the *At-Risk Behavior* corrected (if any)?

safe and at-risk behaviors noted above in the CBI and describes the conversations held with the person being observed. It is required that all at-risk observations must have documented conversations between the participants and provide details of the actions taken to remedy or temporarily make an at-risk behavior safe.

2.5 - Tracking and Trending Observations

Each employee performs at least one observation each month and submits a completed form to the SHE Department for tracking and trending. The data from each of these sections is entered into BAPPTrack® software, a safety data analysis and tracking program.[2] The employee's departmental BBS Steering Team representative reviews the submitted form to ensure quality; if clarification is needed, the observer is contacted to obtain clarification and, as needed, the observation documentation is updated. The BBS Team reports on the number of safe and at-risk observations from the previous month in the monthly safety meeting. Some of these observation's trends have been analyzed and are shown in the results section below.

2.6 - Responding to At-Risk Observations

One of the goals of these observations is for employees to discuss safe behaviors with one another. The discussion's aim to provide certain, positive and prompt feedback on at-risk behaviors with no risk for reporting the observation. This reflects one of the site's guiding principles to respond to at-risk observations, as well (see "Principles of Operation" discussion below). The form is designed to help document how the observer responded positively to and corrected any observed at-risk behaviors. No form is considered complete, of adequate "quality," unless all noted at-risk behaviors have been properly addressed. If an at-risk observation is not adequately documented as closed, a BBS Team member will contact the observer for clarification and ensure a timely and adequate response.

It is important to note that in the beginning of the program, some of the types of observations written down were "condition-based" observations, such as chronic equipment-related issues. Such issues may have been documented to vent an employee's frustration, however, is some of the early non-behavior-related observations that were raised by several employees, the BBS Steering Team decided to address them. This provided the BBS Steering Team with credibility, proving that the team was reading the forms ("listening") and was addressing issues. Sometimes these equipment issues were already being addressed but had been poorly communicated back to the area (the capital approval process takes time). Since then the site's "Work Order" system has been upgraded to allow access by every employee, hence allowing the BBS Steering Team to return to its focus on the behavior-based observations.

Behavior-based at-risk observations include not wearing safety glasses, not wearing hearing protection, or not standing on the ladder properly when repairing an elevated valve. Appropriate at-risk behavior corrections noted for these observations include putting on the safety glasses, inserting the appropriate ear plugs, or stopping the job to evaluate how better to reach the valve under repair. Most of these documented behavior-based observations are easily and immediately corrected.

Recall that the observation form requires that the observer talk to the observed about the at-risk behavior as the observation is being performed. This discussion is the foundation of the BBS program: employees are talking about safety with one another, and hence, at-risk behavior is addressed immediately. In addition, responding to and correcting at-risk behaviors increases the "conduct of operations" term in the overall risk equation, helping to reduce the overall risk and helping to improve the site's safety performance.

2.7 - Communicating Observations to the Site

The BBS Steering Team meets monthly to review the observations and prepares summary reports which are presented in the following month's safety meeting. Since this information competes with other safety meeting topics (company SH&E statistics, incident reviews, etc.), the BBS Steering Team has simplified its monthly report to how many safe as well as how many at-risk observations were noted in the prior month. In addition, the monthly report includes some detail on the responses to correct the at-risk observations. The winners in the "Needle in the Haystack" program (described below) are also noted, as well. Occasionally, the team includes special topics, such as reminders as to who the departmental team members are or refresher charts on how to perform quality observations. The site's quarterly newsletter, the *Cab-O-Sil Courier,* also contains information on the team's efforts on a regular basis, including minutes from recent team meetings and announcements on the winners of the cash prize for finding the "Needle."

2.8 - Addressing the "Human Element"

In the context of this paper, the "Human Element" is defined as the employee-related responses and issues dealing with changing human behavior. Observations by everyone of everyone are key to Tuscola's BBS Observation program's success, with some of the basic human element tenets addressed being 1) everyone on site is involved and responsible, 2) the observations are kept anonymous; and that 3) there are no negative repercussions. The program's positive can-do attitude established credibility. One BBS Steering Team member noted that success occurred because of these program elements reflected in the form shown in Figure 4, as well: concern (awareness), conversation (discussion), ownership and action (change). Stability in the program's expectations by management over the years provided consistency for its sustainability.

An overall goal of these observations is to have employees discuss and then document their safety discussions. One initial issue included keeping a positive attitude on both sides, especially when addressing the communication process (note the steps in the procedure listed in Figure 3). However, some employees simply do not have strong writing skills and struggled to submit "quality" observations. The BBS Steering Team has worked hard to remove the perceived "grading" stigma, and has established a successful peer-review process. This review process provided more visibility for the BBS Steering Team, improved the observer's writing skills through peer review, allowed for coaching by the BBS Steering Team member, and then allowed for re-submittal of the observation, if needed.

2.9 – Searching for the "Needle In The Haystack"

The "Needle in the Haystack" award was created to spice up the observation program in 2007. A $25 gift certificate is given to an employee who observes (finds) one of the two randomly selected "Needles" during the month (the BBS Team selected the Needle from a hat during their meeting). For example, one of the "Needles," unknown to the rest of site for that month, happened to be a production unit shift supervisor. The operator who observed him won the gift certificate when he observed his supervisor during one of the supervisor's daily rounds. As of the time of this paper, 50% of the Needles had been successfully located and observed over the 18-month period. The program has proven to be a fun way for employees to "hunt" for a Needle while performing their observation.

3 - OTHER TUSCOLA RISK REDUCTION EFFORTS

This section briefly describes some of the other safety-related risk reduction efforts at Tuscola, that when combined with those of the BBS Observations, helped improve the site's safety performance. None of these efforts would be successful without the solid and consistent management backing and support of these – and other - efforts over the years. This section begins with a brief description of Cabot's "Drive to Zero" vision, discusses the Tuscola site's "Principles of Operation," and then briefly notes Process Safety Management (PSM), Job Safety Analyses (JSAs), and improved permitting programs.

3.1 – The Drive to Zero

Within Cabot, the "Drive to Zero" initiative is defined as the mindset that will help the company reach its ultimate goal of not hurting anyone. The work elements that Cabot employees can directly control and influence are shown as the foundation of the pyramid shown in Figure 5. This mindset requires a commitment by all who work at Cabot to ensure employees, contractors and visitors are safe within our facilities.

Figure 5. The Cabot "Drive to Zero" Pyramid

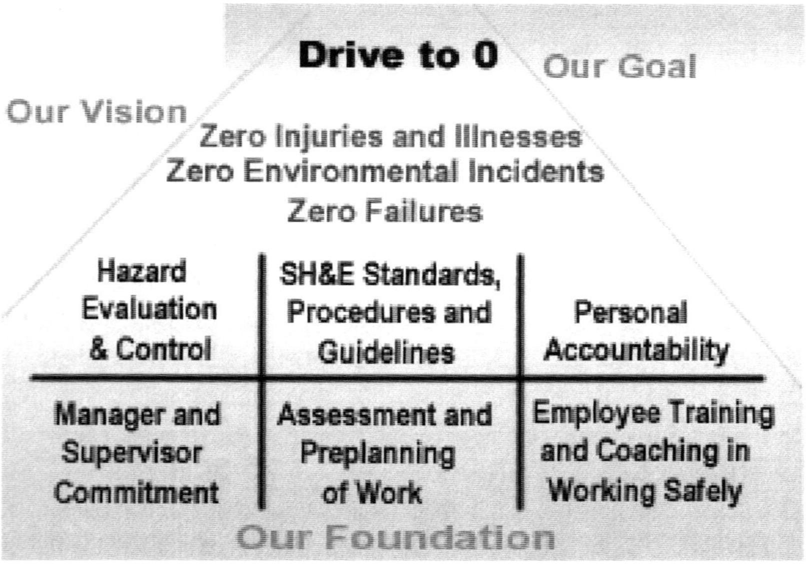

3.2 – Principles of Operation

The Tuscola site's "Principles of Operation" is a code of conduct aimed to help achieve injury free and incident free performance. These principles were issued in 2008, summarizing potential high risk behaviors and listing expectations that must always be followed. The code formally adopted elements and re-emphasized existing safety expectations and rules.

The Principles are as follows:

- Always operate equipment and processes within designed safety and environmental limits.
- Always move to a safe controlled condition OR location when a situation is not understood.
- Always seek assistance and involve people with expertise in changes or improvements that affect equipment or procedures.
- Always understand and address deviations and ensure proper corrective actions are implemented AND communicated.
- Always report environmental or safety deviations promptly and accurately.
- Always follow safe work practices and written procedures and act to stop unsafe conditions and behaviors.
- Always address abnormal conditions and clarify / understand all procedures before proceeding.
- Always use tools/equipment/processes for their intended use. Never compromise any safety systems and/or interlocks.

These Principles are currently posted across the site.

3.3 - Process Safety Management

One of the Process Safety Management (PSM) elements is "employee involvement," or what can be referred to as the "human element" of PSM. Employee involvement is an essential underlying element: humans are the designers, fabricators, operators, maintainers and managers of our processes. Cabot has issued four corporate PSM Standards to date on the following PSM elements: Process Safety Information (PSI), Process Hazards Analysis (PHA), Management of Change (MOC) and Pre-Startup Safety Reviews (PSSR). The Tuscola site is covered by the OSHA PSM standard and has been actively identifying and addressing both the OSHA and Cabot standards.

3.5 – Job Safety Analysis (JSA) Reviews

Job Safety Analysis reviews are performed with a member of management every weekday, following a monthly schedule for each member of management to perform at least one a month. The JSA reviewer provides a third set of eyes on the jobs being performed. The site has a dedicated JSA approver, as well, who follows many of the jobs during the course of the day.

3.6 – Contractor Safety Programs and Involvement

It is a Cabot expectation that all contractors follow all safety rules, as well. Contractors attend their own special monthly safety meeting run by Tuscola personnel on site, focusing on contractor issues and ending with lunch. In addition, one contractor has informally noted its use of the Tuscola BBS Observation ideas to help improve safety discussions between its employees and with their Cabot hosts.

4 - OVERALL TUSCOLA SAFETY METRICS AND RESULTS

This section describes some of the improvements and results in the site's safety performance over the last decade. These improvements are a result of the combination of safety efforts described above, including the discussions that occur between employees while performing their BBS observations. Recall that the goal of these efforts is to prevent injuries to personnel, damage to facilities and damage to the environment by understanding the risks and reducing them as much as is practical. The metrics noted below have been established based on standard industry practices and help support evidence that these risk reduction efforts are making a difference.

The trends tracked through the BST® software are described first. Then the metrics from our personnel injury-related and our facilities-related (process hazards) incidents are described below. The BBS observations became mandatory in 2004 and are referred to in the charts as "FY04" (the Cabot "fiscal year," FY, where the FY runs from October to September).

4.1 - Behavior Based Safety (BBS) Observation Metrics

As was noted above, the BBS observations became mandatory for all employees in 2004: every employee must perform, document, and submit at least one observation each month. The current observation form is shown in Figure 4, where Safe or At-Risk behaviors for the person's body, the person's procedure or the person's personal protective equipment (PPE) are noted. The following charts represent the data that has been entered over the last decade into the BST® software. Although the observation form has evolved over this time period, it is hoped that the general trends shown below are evident of the site's success when employees talk frequently about safety.

As is shown in Figure 6, the number of safe observations documented for the site increased from 10- to 15-thousand per year before FY04 to almost 25-thousand per year in FY08. This is a favorable trend, as this indicates that over this time period employees are observing and talking more about their safe behaviors. As is shown in Figure 7, the number of at-risk observations for the same period has decreased from more than 300 per year before FY04 to less than 200 per year in FY08. Again, this is a favorable trend, as this indicates that employees are recognizing, observing, and talking more about at-risk behaviors. In addition, the observation procedure requires that the observer note how the at-risk behavior was then addressed and corrected.

Recall that the observation form in Figure 4 has evolved during this time period, as well. Some of the form upgrades included adding more line items as check boxes, hence, providing the observer with more options to check for either Safe or At-Risk behavior. Adding more check boxes only supports and reinforces the favorable trends noted above: an increase in Safe observations corresponds to more discussions between employees about safe behavior (Figure 6); the decrease in At-Risk observations reinforces fewer at-risk behaviors (Figure 7).

4.2 - Personnel Injury Safety-Related Metrics

The personnel injury safety-related metrics are based on the basic definitions of the type of injury and are classified as "First Aid," "Medical Treatment" or "Lost Time." Looking at the risk equation, the consequences of a first aid injury are small and essentially reversible when compared to the consequences of a medical treatment injury, which may be a potentially irreversible injury and may be designated as "lost time." As is seen at the top of the safety pyramid (Figure 1), the worst irreversible injury is death. "First Aid" cases are defined as incidents where employees are exposed to materials or process hazards that harm them, such as minor scratches, minor cuts, thermal or chemical burns, or splinters. In addition, first aid cases do not prevent the employee from returning to and performing their normal work duties. Medical treatment cases, however, require more medical professional intervention, and may be severe enough to prevent the employee from returning to work (a "lost work day" case)." Within the medical treatment definition are treatments that require prescription medications (antibiotics, pain killers), surgical procedures (staples, stitches, removal of embedded foreign bodies), fractures or broken bones, loss of consciousness and hearing loss.

Figure 6. The Trend of Documented Safe Behaviors

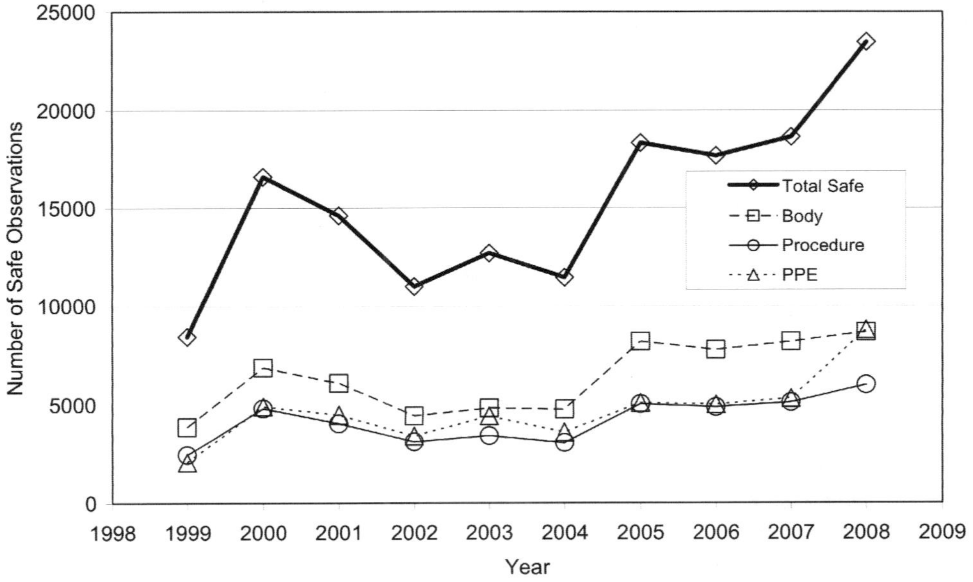

Figure 7. The Trend of Documented At-Risk Behavior

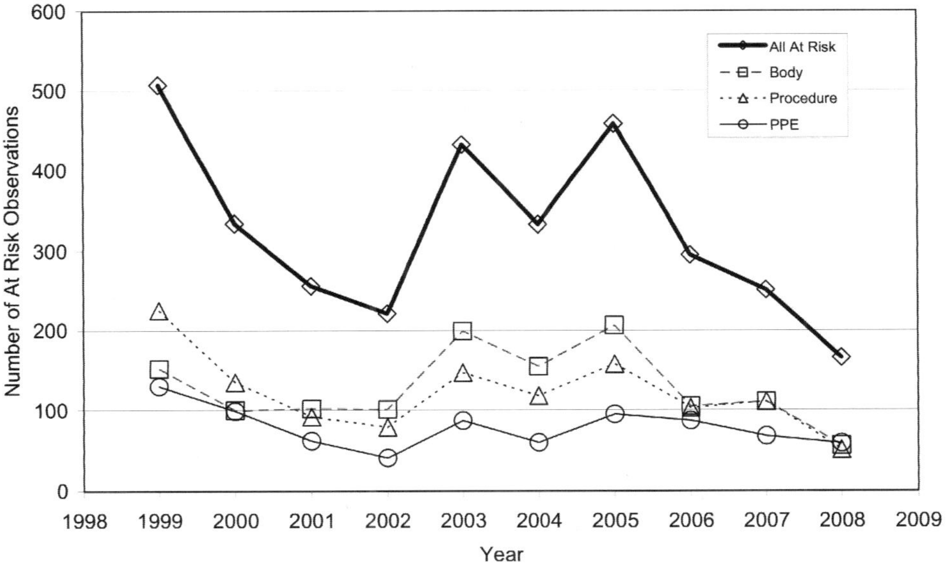

Tuscola's personnel safety-related metrics are shown in Figure 8. The trends are favorable. There were six first aids in 2008, less than the fifteen in 2007. There were no medical treatment or lost day injuries in 2008 and there have been no lost day injuries in the last three years.

4.3 - Facility Safety-Related Metrics

The facility safety-related metrics are based on the basic definitions of the type of process incident and on its relative consequence. New corporate definitions of Near Misses and Process Safety Events (PSE) were established in 2005. PSEs include significant fires, explosions, deflagrations, and toxic releases. PSE incidents with hazardous materials also include vessel overfilling, major leaks or releases, rupture or failure of process equipment, and process upset events that open up atmospheric relief systems (the last line of defense protecting equipment). Near misses include incidents that could have been severe if they had occurred at a different time or at a different location on site. Property damage includes all incidents in low hazard areas, as well, such as those that occurred in packing and shipping.

Tuscola's facility process hazard-related metrics are shown in Figure 9. There is a favorable downward trend of number of reported property damage and process safety events. The upward trend of near misses may indicate greater employee awareness of hazardous conditions or potentially hazardous events and, hence, proactive reporting and addressing them. These trends reinforce employee's sharing of low consequence events (better reporting) and more efforts to attack the base of the Safety Pyramid (Figure 1).

5 - CONCLUSION

Reducing risk by improving everyone's "conduct of operations" and operational discipline requires the awareness, knowledge, and commitment of everyone.[3, 4, 5] The Behavior Based Safety Observation program that has evolved over the last decade at the Tuscola FMO site has contributed, in part, to safety-related discussions that have improved safety awareness, have helped share safety knowledge, and have helped sustain commitments to safe behavior. Everyone on site is involved and responsible. They understand that their observations are documented as anonymous, there are no negative repercussions, and they have experienced a consistent management support, approach and expectation year after year. The success of the BBS Observation program has helped raised the denominator in the risk equation and helped reduce overall personnel- and process-related risks.

Figure 8. The Personnel Injuries-Related Incident Trend

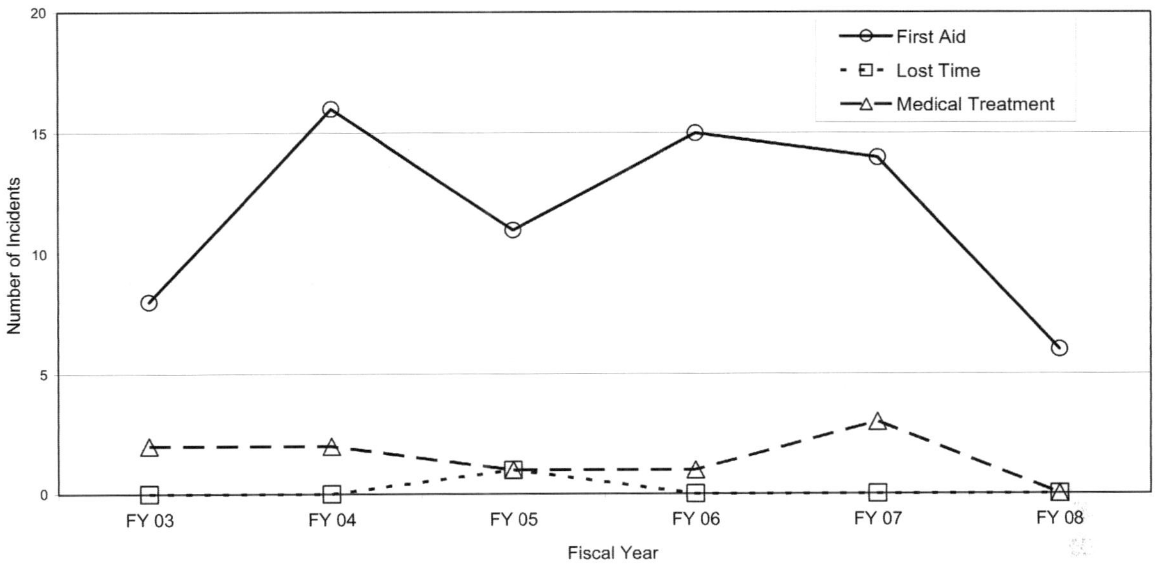

Figure 9. The Process Hazards and Facility-Related Incidents Trend

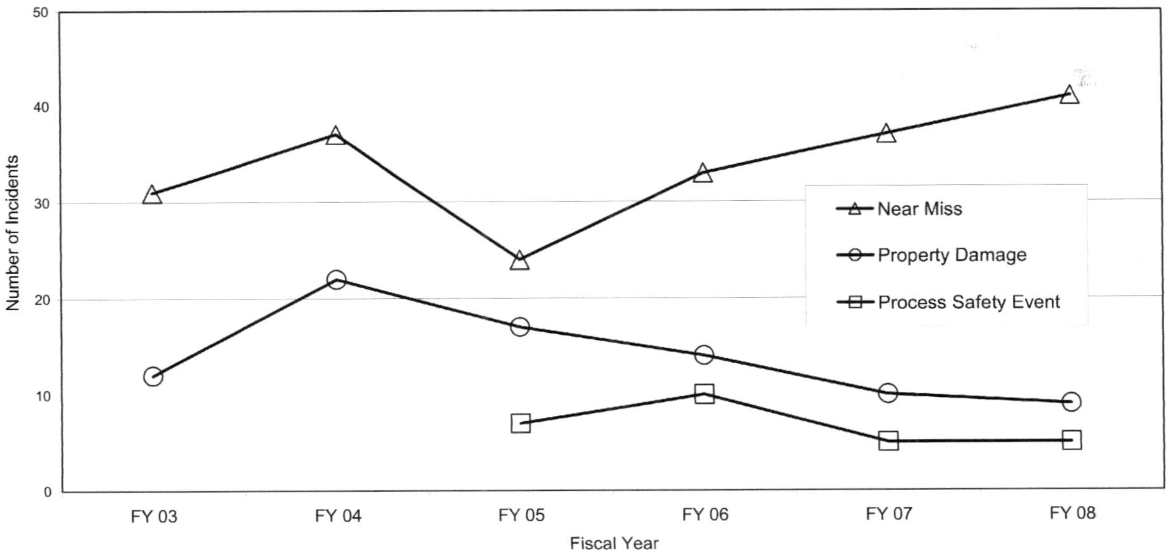

6 - FUTURE PLANS

The ultimate goal of the Behavior Based Safety Observation program is to eliminate the use of the observation sheets: the behavior will evolve from employees asking to observe one another to employees asking to be observed by one another. As behaviors improve, habits improve, operational discipline improves, the conduct of operations improves and the overall risk is reduced. In addition, these discussions between employees may help reinforce behavior-related improvements in other areas as well. These other areas include environmental- or quality-related incident reductions, such as reducing spills and their impact, eliminating environmental non-compliances, reducing customer complaints, and eliminated off-quality product as a result of the improved operating discipline of each and every employee.

References

1. The "Critical Behavior Index (CBI)" is a term used in the BST® Training and in its BAPPTrack® software (Reference 2).

2. BST® BAPPTrack® software, Version 2.x, © 2008, Behavioral Science Technology, Inc., Ojai, California 93023 (805) 646-0166.

3. Klein, James A., and Bruce K. Vaughen (2007), "Evaluating and Improving Operational Discipline," AIChE, *Plant Process Safety Symposium*, 22-26.

4. Klein, James A., and Bruce K. Vaughen (2008), "A Revised Program for Operational Discipline," AIChE, *Process Safety Progress*, Vol. 27, No. 1, March, pp. 58-65.

5. Williams, Joshua H. (2008), "Employee Engagement; Improving Participation in Safety," ASEE, *Professional Safety*, December, pp. 40 - 45.

Safety Culture Implementation Programs in Developing Countries

Lopes Alves and Luiz Fernando Oliveira (DNV Brazil)

Ron Hallmark and Robin Pitblado (DNV Houston)

ABSTRACT

The Texas City Refinery disaster and subsequent Baker Panel Report have been driving factors for US companies to focus on Safety Culture programs and the Center for Chemical Process Safety (CCPS) is presently drafting a book on Operational Discipline and Conduct of Operations. These types of initiatives are not only taking place in the US but also in developing countries. This paper contains 3 examples of culture improvement, two in Brazil and one in Pakistan. These examples illustrate that, as expected, there are many similarities in safety culture development and implementation between these developing countries and the USA, but also some differences. This paper highlights successful safety culture examples from both Brazil and Pakistan in the context of development of safety culture in developing countries.

INTRODUCTION

Structured safety management systems and risk assessment programs have been important drivers for safety improvements in the past 20 years. However these have been more successful for occupational safety than for process safety as can be seen in the comparison of trends.

Figure 1. Comparison of Occupational and Process Safety Attainment
Source: DNV Research

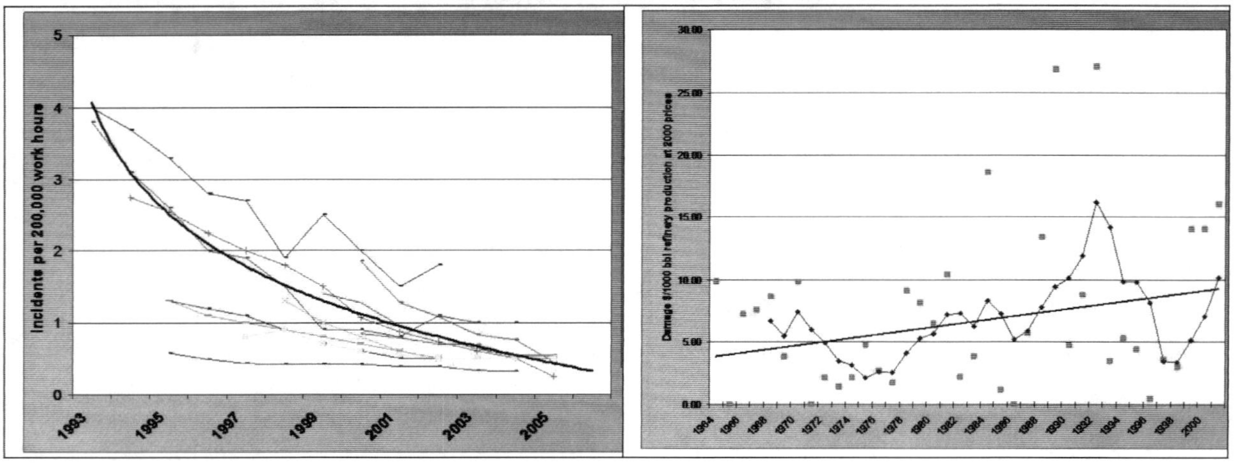

The Texas City vapor cloud explosion is only the latest in a series of major accidents globally (Toulouse ammonium nitrate, Buncefield tank explosion, Antwerp pipeline) that reinforce the finding of Wharton Business School (Kleindorfer, 2007) that EPA RMP statistics show no decline in major accidents. Clearly, what works for occupational safety is insufficient to drive down process safety incidents to anything like the same degree. Extra efforts are necessary and these will be a mixture of enhanced technical integrity programs, reinvigoration of safety management systems (SMS), and deployment of proactive safety culture programs. The Baker Panel (2007), which had a charter to examine SMS and safety culture at BP's North American refineries made several strong recommendations in both these areas.

While the USA has a number of safety culture or behavior change programs deployed for many years at some companies, experiences overseas are less well documented.

Methodologies for culture and behavior improvement have two main academic backgrounds, psychology and social anthropology, and these lead to somewhat different solutions. Programs at site level tend to be implemented by engineers and supervisors and this makes also for some difficulties in communication. A good survey of current safety culture theory is provided in (Choudhry et al, 2007). This highlights that there is not yet a fully accepted definition of safety culture. DNV prefer a definition such as:

> "Safety Culture is the common set of behaviors and norms which are set and maintained by internal and external motivation forces that affect safe system performance at all levels in a facility or the wider company. Proactive sustainable programs can drive behavioral change to enhance Safety Culture".

Our rationale for preferring this definition is that it recognizes that culture is equally important at all organizational levels of a company and is affected by both internal and external drivers. This is important in developing countries where the local traditions can be as strong or stronger than the company traditions. It also confirms that safety culture and behavior improvement can be managed for positive improvement through the implementation of a long term program. A good summary of current thinking on culture programs is provided in the literature in Table 1.

Table 1. Elements of Safety Culture Programs from Vecchio-Sudus and Griffiths (2004)

Changing attitudes and behaviors: Safe behaviors can be enhanced by capitalizing on activities such as verbal instructions, training, and warning signs. Long-term values include employees being able to work without injury so they can continue to provide earning for both the company and for her/his family.
Management commitment: Management plays a key role in promoting a positive safety culture. This can be best demonstrated by allocating resources, time, *walk the talk*, inspections, by participating in risk assessments and consultative committee meetings, and by completing actions.
Employee involvement: For a positive safety culture, employees' involvement, ownership and commitment is necessary; in particular empowerment promotes feelings of selfworth, belonging and value.
Promotional strategies:
(i) Mission statements, slogans and logos;
(ii) Publish materials (library, statistics, newsletters);
(iii) Media (posters, displays, audiovisual, e-mail, Internet).
Training and Seminars: Training activities should include short talks, group meetings, training for personal fitness, hygiene, workplace stress and responsibilities towards safety (including compliance with rules and regulations, hazard identification and risk assessment, incident investigation and job safety analysis);
Special campaigns: This item should include Health and Safety Week, health promotion, safety inductions, emergency response, incident reporting and investigation, risk assessment, introduction to existing health, safety and environment management systems.

DNV has evolved a set of common factors for development of Safety Culture based on the professional literature and work experience listed below. The delivery process matches well with a similar system implemented within a major oil company in the Netherlands.

Table 2. Principles for Culture Enhancement Programs

Delivery Principles for Projects	1. Ownership – full participation by the client management and workforce 2. Eclectic approach – mixture of behavior ideas (punish / reward – DuPont and BST) and culture systems (management programs). Where possible link closely to safety management system and risk based programs. 3. Follow-up / improvement
Process – 4 steps	Like the consulting model Identify/ Diagnosis, Develop solution, Implement, Sustain
Key elements of culture survey always to address	Competence, Collaboration, Conflicting goals, Compliance Incentives, Org Learning, Risk awareness, Error tolerance

These principles are demonstrated by the three case studies provided below.

CASE 1 – BRAZIL MINING EXAMPLE

The first example is a large multi-national mining company with dozens of operating sites in Brazil. The company's Senior Management felt that with their expansion and listing on the New York Stock exchange, reputation issues required them to upgrade their HSE performance to "Industry Best" as well as develop their own unique safety culture. Company management decided that they wanted a full Safety Culture program to be integrated into a new risk-based Safety Management System. DNV was contracted for the project to simultaneously develop and implement safety management systems and behavioral change. This project has been a collaborative effort between DNV and the Mining Company with DNV providing the initial plan, coaching, training, and oversight. The Company directed that the project would be conducted at four pilot sites, which combined have several thousand employees and contractors. They further stipulated that as well as these pilots, certain safety programs were to be applied at all of their Brazilian sites.

The project methodology used by DNV involved a 4 stage approach: Diagnosis, Development, Implementation and Sustainability.

Diagnosis – Establish current safety culture norms using 3 different tools: culture questionnaire, 1-on-1 interviews, and focus group discussions. The company presented DNV with 17 SMS Element topics to address but with wide scope to build these out for maximum effect. DNV developed a fit for purpose gap analysis protocol which was utilized to gauge the safety management systems in place.

Development – DNV drafted a master project plan which included mapping a solution and selection of company champions, required training of Leaders and Champions, and coaching to be provided to ensure quality implementation. The project required deployment of a team of over 50 professionals – psychologies, social anthropologists, medical doctors, and engineers. Some of those are embedded at the 4 pilot site locations providing implementation assistance. DNV is also developing solutions and providing specialist services to all the sites.

Implementation – As part of the implementation, all managers, supervisors, and HSE staff were trained in Risk awareness which included both theory and practical application of hazard identification and hazard control techniques. These same groups were also trained in Behavioral Change processes, particularly in behavioral observation techniques. A behavioral observation program was then proposed and implementation was initiated.

Sustainability – Several phases of the project have been completed such as Champion training, risk awareness and behavioral change training. However, critical task analysis which include check-lists are still on-going. DNV is in the process of training company auditors who will conduct internal audits after DNV role is finished. Performance measures already indicate substantial improvement. The company's management is committed to achieving and sustaining industry best HSE performance with a unique Safety Culture.

The risk competence wheel below in Figure 1 illustrates the concepts that have been successfully used in the Brazilian Mining project. Note that Leadership's active support drives risk perception/acceptance, safety knowledge and skill (competence), internal motivation, and finally operational discipline.

Figure 2. Risk Competence Wheel Model

CASE 2 – ELECTRICITY TRANSMISSION COMPANY BRAZIL

DNV in conjunction with the local university executed a culture improvement project for a major electricity transmission company that was experiencing a high number of serious accidents amongst its line workers. This motivated the company to seek an improvement process. Accidents were related to falls or electrocutions. They identified that the main cause of the serious accidents were either poor job procedures or lack of compliance with them.

The solution involved a multi-faceted approach – enhanced safety management system, risk assessment, safety culture diagnosis, procedure review and rewriting, training in new procedures, and task observation.

A detailed safety culture questionnaire-based was undertaken asking respondents to rate 25 important characteristics from low to strong influence. Many questions showed high consistency between all staff and all locations, other showed strong bifurcation (e.g. safety takes priority over production) where 2/3rd strongly agreed and 1/3rd strongly disagreed. Inherent to questionnaires, only a few questions lead to a definitive picture. Instead the questions have greater meaning when movements in the scores are compared to a resurvey in 2 years time. This matches mining project experience where focus group interviews alone delivered as much directly useful data as the questionnaires and 1-on-1 interviews combined, at lower cost.

To underpin the safety management system activity a baseline survey was carried out at multiple field locations as well as head office. This involved interviews with executives, managers, supervisors, and the workforce.

Procedures were rewritten by a local university, UFRGS, and company supervisors working as a team and reviewed by DNV. Procedures for update were selected on a risk-basis linking these to past incidents. They used an effective format that enhanced clarity and simplified procedures to essentials, expressed clearly and using photographs where helpful, and highlighted safety issues explicitly. Standard symbols (colored hardhats) were used to differentiate roles where multiple parties were involved. The procedure sheets all had a common format making for easy use. Key sections included:

- Tools required
- PPE required
- Risks
- Safeguards
- Prior tasks to be completed first (referenced procedures)

Task observers were trained, and a major increase in formal observations was initiated. These started at 200 per year yr and in the following two years covered 2029 and 2529 observations. The key basis for the observations included:

- Use of a specialist team
- Everyday is an Observation Day
- Observation focuses on critical tasks
- Specific indicators for the function
- Group evaluation based on its performance
- Detection and treatment of specific and systemic causes
- Supervisor is responsible for results
- Zero tolerance for risk assumption
- 100% emphasis on the use of the new Procedures Manual

A gauge of the success of the program is provided in the table below. The percentage of correct use of procedures increased over a 9 month period from 20% to 79%. This was linked as seen to supervisor support workshops and to a revision to the manual where it was shown to be unclear or inefficient.

Table 3. Improvement in Task Execution linked to task observation

Month	Main Actions	Procedures Manual Rules correctly followed
Oct/04	Start – specialist team	20%
Nov/04	General Safety Alert	30%
Dec/04	Action Plan for the Supervisors	40%
Jan/05	Monthly Supervisor's Ranking	47%
Mar/05	Task Observation Workshop	60%
Abr/05	Review of the Electrician's Manual	63%
May/05	First 100% standard group	56%
Jun/05	Supervisor's Workshop	69%
Jul/05	Second 100% standard group	79%

There was almost immediate success. The safety management system was reviewed and given heightened awareness especially among processes related to risk assessment. The improved/enhanced SMS also assisted in the management and control of contractors. The task observation program supported the procedure compliance. Figure 3 below presents the results of a single year of task observation efforts. It is interesting to observe the decline in accidents concurrent with increased compliance to the procedures.

Figure 3. Observed relationship between accident rates and compliance with job procedures

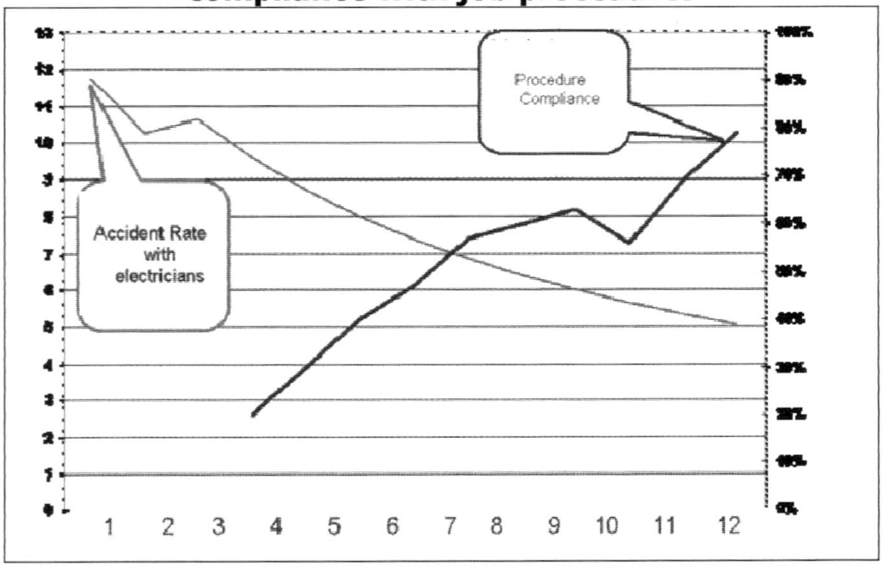

The results so far from the Safety Culture project have been improved accident rate, compliance with procedures and a major change in Safety Culture.

CASE 3 – PAKISTAN EXAMPLE

One of the co-authors of this paper, Ron Hallmark, spent 2 years working in Pakistan as an HSE Manager for a multinational Oil/Gas Company. His task was to transform the oil/gas asset into a top-tier HSE performer. Pakistan had many similarities to Brazil in that the government had promulgated strict health & safety regulations. However, there was little effort to enforce the regulation. Additionally, Pakistan lacked a pro-active industry organization similar to Brazil's. The Pakistan workplace culture is typical of many developing countries. It is very authoritarian, with a sense of honor, respect for elders and persons in higher authority. This can lead to overruling of safety procedures if it is not carefully managed. This is similar to Brazil where large industry had been nationalized until recent decades.

Paralleling the methodology used in Brazil, the Pakistan approach also included 4 stages which map to the structure provided earlier. Using the same 4 factors:

Diagnosis – A Safety Culture questionnaire was conducted as well as a gap analysis vs. the Parent Oil/Gas Company SMS.

Development – Set up a risk based master project plan which included selection of Champions for the SMS elements, variety of training for managers and Champions and workers. The SMS elements were prioritized based upon risk .

Implementation – As part of the implementation, all Managers, supervisors, and HSE staff were trained in SMS, the expectations of each SMS element, the intended results, and individual goals and objectives for managers and workers.

Sustainability – The Parent Oil/Gas Company set up an audit schedule to ensure compliance with the SMS. Quarterly SMS reports were required by corporate as well as an annual letter of HSE compliance.

Pakistan Comments/Summary

Some key points learned from the experience related to behavioral change:

- Management commitment - Pakistan management, once convinced of their goal to achieve industry best HSE, played a key role in promoting a positive safety culture by providing the necessary resources, becoming actively involved in HSE activities such as HSE audits, setting up of HSE Committees, setting HSE targets for all levels of management, leading safety meetings, making presentations at safety emphasis campaigns, conducting safety observation audits, and ensured safety actions were completed via regular reporting.

- Employee involvement - Employees were empowered to openly report hazards and to stop work where it was deemed hazardous. Employees were involved in development of a safety training matrix, approval of PPE by testing specific types of PPE and ensuring they were fit for purpose before requiring that particular PPE.
- Behavioral Observation Program - Once the concept of just culture was explained; employees readily accepted the observations conducted by managers, staff, and supervisors. They especially enjoyed the positive dialogues and rewards instilled for safe behavior. The workers came to appreciate the No-Blame/error tolerance culture.
- Collaborative Efforts - Employees were solicited for input for the new HSE policy statement, employee committees proposed slogans and topics for safety campaigns, contests were held amongst employees for best safety posters and slogan for the month, a safety newsletter bulletin was published by HSE staff with assistance from workers.
- Competency & Training - Training plan included cascaded training, from Classroom training to team safety meetings which were planned and scheduled, topics were generally proposed by HSE Team and were based on risk or lessons learned. Classroom communication techniques had to be revamped to include more 2-way dialogue and team exercises. All training matrices included mandatory risk awareness training.
- Special campaigns - Special campaigns were conducted based upon the highest identified hazards (8-10 highest rate of incidents) and later upon lessons learned and results of data from inspections and audits. When the SMS was introduced, special campaign was introduced for roll-out.

PAPER CONCLUSION AND SUMMARY

There are similarities and differences in safety culture programs in developing countries compared to the USA and Europe. The main difference lies in the lesser regulatory regimes – either through an absence of regulation or a lack of enforcement.

- There are societal norms which differ, such as a greater respect for authority and willingness to follow instructions from managers. This can be a strength of the management system promulgates for full safety program not driven by temporary operational demands, but also a weakness.
- Many developing companies have a command structure where reward and punishment is a driving behavior factor.

- The programs DNV has been involved with tend to be holistic involving the safety management system, risk assessment and safety culture / behavior programs.
- These have followed a common 4 tier structure consisting of diagnosis, development, implementation and sustainability.
- These programs follow international norms and are a mixture of classical behavior approaches with a more management system culture approach.
- In the USA, such holistic programs are unusual as PSM regulations have already driven the SMS and risk assessment elements and thus the culture side is often executed as a separate program, but in that PSM environment.

REFERENCES

1. Baker III James (2007) The Baker Panel: BP U.S. Refineries Independent Safety Review Panel Report, 374pp, January, 2007.

2. Choudhry RM, Fang D, Mohamed S (2007) The nature of safety culture: A survey of the state-of-the-art, Safety Science 45, 993–1012.

3. Hudson P (2007), Implementing a safety culture in a major multi-national, Safety Science 45, 697–722.

4. Kleindorfer P (2007) Accident Epidemiology and the RMP Rule: Learning from a Decade of Accident History Data for the U.S. Chemical Industry, Wharton Business School for Office of Emergency Management EPA December 18, 2007.

5. Vecchio-Sudus, A.M., Griffths, S., (2004). Marketing strategies for enhancing safety culture. Safety Science 42, 601–619.

New Insights into Computing PFDavg for Pressure Relief Valves Based on Proof Test Data

Julia V. Bukowski, Villanova University, Villanova, PA 19085, USA, julia.bukowski@villanova.edu

William M. Goble, exida, 64 N. Main St., Sellersville, PA 18960, USA, wgoble@exida.com

ABSTRACT

Previously we analyzed several PERD[1] Pressure Relief Valve (PRV) proof test data sets using Quantal Response Analysis (QRA) and obtained estimates for λ_D, the constant failure rate for the "stuck-shut" failure mode, in the range of $[10^{-8}, 10^{-7}]$ failures/hour. We also found that in that range of values, λ_D alone explained only a small fraction of the observed failures; the remaining failures were modeled by probability of initial failure (PIF) in the range of 1% - 1.6%.

Using Statistical Signature Analysis (SSA), we now re-analyze two of the data sets, this time considering more complex models of $\lambda_D(t)$ to explain the observed failures and to compute average probability of failure on demand (PFDavg). Our findings are:

- Any analysis of proof test data, based solely on operating times until proof test and results of proof test (pass/fail), must necessarily admit multiple explanations for the failure mechanisms underlying the test results, i.e., one data set will be statistically consistent with more than one $\lambda_D(t)$.
- Since PFDavg serves to measure risk reduction, it should include all dangerous failures regardless of underlying cause. By using more complex models for $\lambda_D(t)$ than are currently considered, e.g., models which can include constant λ_D, both zero or non-zero PIF, and models for infant mortality and/or failure rate growth, it is possible for $\lambda_D(t)$ to account for all observed failures regardless of underlying cause.
- Each different explanation of failures observed, i.e., each different $\lambda_D(t)$ model produced from a single data set, potentially produces a different value for PFDavg for any given proof test interval, T_P. Thus, one data set can support multiple values of PFDavg for any fixed T_P. However, *for the data analyzed to date* and for a fixed value of λ_D, these difference are small for $T_P > 4.5$ years. Generally, in these cases, PFDavg > 0.01. Hence, once T_P reaches about 4.5 years, the data analyzed support Safety Integrity Level (SIL) 1 behavior for all of the complex models of $\lambda_D(t)$ explored thus far. For values of $T_P < 2$ years, *some* complex models for $\lambda_D(t)$ support SIL 2 behavior. But we can not tell by currently available analysis methods which models best represent the actual failure mechanisms responsible for the proof test data observed. Thus, in general, current data supports SIL 1 behavior for PRV.

[1] See Acknowledgement for contact information for and additional description of the CCPS PERD Initiative.

INTRODUCTION

In order to assign a SIL to equipment in low demand applications, we must be able to compute the average probability of failure on demand (PFDavg). To compute PFDavg, we must first have a model for $\lambda_D(t)$, the failure rate of the equipment in the dangerous failure mode. A dangerous failure occurs when equipment designed for prevention or mitigation of an unsafe condition cannot properly respond to the unsafe condition, i.e., the equipment fails on demand. For example, consider a pressure relief valve (PRV). In normal operation, the PRV is closed. Should it fail in the "stuck-shut" mode, it would be in a state of dangerous failure as it would be unable to respond to an overpressure event if one occurred.

Once we have determined a model for $\lambda_D(t)$, we can compute PFDavg which is defined as

$$\text{PFDavg} = (1/T_P) \int_0^{T_P} \text{PFD}(t)\,dt = (1/T_P) \int_0^{T_P} \{1 - R(0) \exp[- \int_0^{t} \lambda_D(s)\,ds]\}\,dt. \quad (1)$$

Current practice for computing PFDavg is based on three assumptions: $R(0) = 1$, i.e., that each PRV is working at $t = 0$, $\lambda_D(t)$ is adequately characterized by a constant, λ_D, and proof testing results in perfect restoration. Based on these assumptions, PFDavg is usually approximated as

$$\text{PFDavg} \approx \lambda_D * T_P/2. \quad (2)$$

Does currently available evidence support the appropriateness of these assumptions? No. For PRV, the generally accepted range of values for λ_D is 10^{-8} to 10^{-7} failures/hour. (For some types of valves, some in the European safety community advocate even smaller values for λ_D, e.g., [1].) However, even if we allow λ_D to be as large as 10^{-7}, we find that λ_D accounts for only a small fraction of the total number of stuck-shut failures observed in available PRV proof test data sets [2].

While retaining λ_D as a constant, in [2, 3] Bukowski and Goble relaxed the assumption that $R(0) = 1$ to $R(0) \leq 1$ by introducing the Probability of Initial Failure (PIF), defined as $\text{PIF} = 1 - R(0)$, and showed that the remaining observed failures not accounted for by λ_D can be explained by values of PIF ranging from 1% - 1.6%. With this modification, PFDavg can be approximated as [3]

$$\text{PFDavg} \approx \text{PIF} + (1 - \text{PIF}) * \lambda_D * T_P/2. \quad (3)$$

Note that (3) reduces to (2) if PIF = 0, i.e., if $R(0) = 1$, so (3) includes (2) as a special case. In [3] it was also shown that even if λ_D is as large as 10^{-7}, PIF dominates PFDavg. This approach treats all failures not accounted for by λ_D as initial failures and provides a conservative (though somewhat controversial) approach to the assessment of PFDavg.

In [4, 5], using a new method of statistical analysis, Bukowski and Goble showed that more than one combination of λ_D and PIF were statistically consistent with the proof test data and that different combinations lead to different PFDavg assessments. They did not, however, explain why these different combinations existed and why proof test data supported more than one explanation for the underlying failure mechanisms responsible for the observed failures. Understanding this is important to being able to deduce which of the many possible combinations of λ_D and PIF best explains the observed proof test data. Lastly, they suggested that more complicated $\lambda_D(t)$ be considered when trying to explain the observed failures.

This paper:

- argues that, logically, the calculation of PFDavg (from which SIL is assigned) must account for all dangerous failures regardless of underlying cause.;
- explains how $\lambda_D(t)$ together with PIF can model all dangerous failures regardless of underlying cause.;
- underscores why seeking to identify $\lambda_D(t)$ and PIF based on proof test data alone necessarily results in multiple $\lambda_D(t)$ and PIF combinations which provide equally good explanations for the observed proof test data;
- compares and contrasts two methods for the statistical analysis of proof test data;
- revisits two proof test data sets previously analyzed by Quantal Response Analysis (QRA) and re-analyzes the data using the new Statistical Signature Analysis (SSA) method now including both constant and declining failure rates in $\lambda_D(t)$ as well as both zero and non-zero values of PIF;
- demonstrates that the different $\lambda_D(t)$ and PIF combinations resulting from SSA produce different PFDavg for the same proof test data;
- stresses the importance of appropriate data collection, analysis, and sharing of results to reduce costs while maintaining or improving safety.

NOTATION

CCPS	Center for Chemical Process Safety
CPI	chemical process industries
FMEDA	failure mode, effects, and diagnostic analysis
FITS	10^{-9} failures/hr
HPI	hydrocarbon process industries
PERD	Process Equipment Reliability Database
PFDavg	probability of failure on demand averaged over $[0, T_P]$
PFD(t)	probability of failure on demand as a function of time
PIF	probability of initial failure; 1-R(0)

PRV	pressure relief valve(s)
QRA	quantal response analysis
R(0)	initial reliability; 1 - PIF; may be less than 1
SIL	safety integrity level(s)
SSA	statistical signature analysis
T_{IM}	end of the period of infant mortality failure rate
T_P	length of a particular PRV time in-service until proof test
λ_D	constant failure rate of the dangerous ("stuck-shut") failure mode
$\lambda_D(t)$	failure rate of the dangerous ("stuck-shut") failure mode as a function of time; may be a constant

USING $\lambda_D(t)$ TO MODEL ALL DANGEROUS FAILURES

Some safety analysts believe that it is necessary to include only certain failures resulting from some particular types of failure mechanisms in the computation of PFDavg. Often they use the PFDavg approximation given in (2) to justify including only failures attributable to the constant failure rate, λ_D. We believe it is self-evident that, from a safety standpoint, a dangerous failure is a dangerous failure regardless of cause and must be accounted for in PFDavg. After all, if a PRV does not respond to an overpressure event, the safety consequences exist regardless of why the dangerous failure occurred. Therefore, the question is not which dangerous failures to include in PFDavg but how to include them.

Consider the case where extensive *time-to-failure* data exists for a type of safety equipment. (This is not the case for the "stuck shut" mode in PRV but we will consider that complication subsequently.) Say we have 1000 equipment units and for each unit we know the entire state history over an interval of 0 to 5 years. Further assume that in the 5 year period exactly 7 units (0.7%) fail and the remainder are successful for the entire 5 year period. Figures 1 - 3 show three possible sets of time lines for 10 such units: 3 of the 993 successful units and all 7 of the failed units. The solid lines indicate time intervals when the unit was in a state of success and the F lines indicate time intervals when the unit was in a state of failure. If a unit is successful over the entire 5 year interval, its solid line ends in the letter S. The time-to-failure patterns are quite different in each of the figures. Figure 1 shows 5 of 7 failures clustered in the first year. Figure 2 shows the 7 failures spread over the entire 5 year interval. Figure 3 shows 6 of 7 failures clustered in the last year. Clearly, the three figures show different failure patterns that would be represented by three different types of $\lambda_D(t)$ which we could estimate given the time-to-failure data.

For example, Figure 1 would likely be consistent with a declining failure rate; Figure 2, with a constant failure rate; and Figure 3, with an increasing failure rate. These three examples are deliberately simple, each exhibiting a single failure rate characteristic. In general, time-to-failure data which included many units over an extended observation

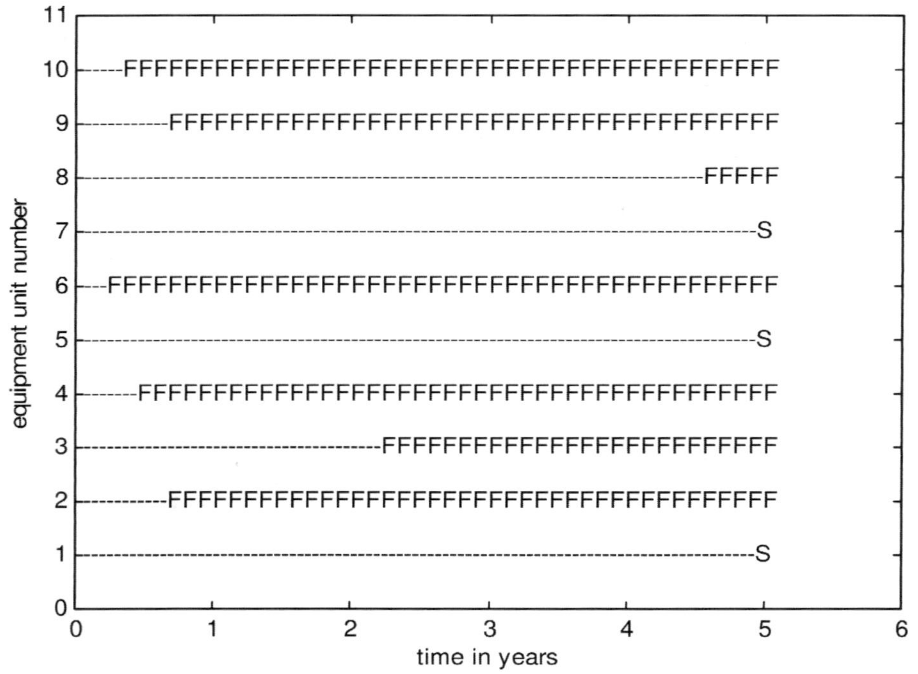

Figure 1. Example Time-to-Failures Clustered in First Year of Service

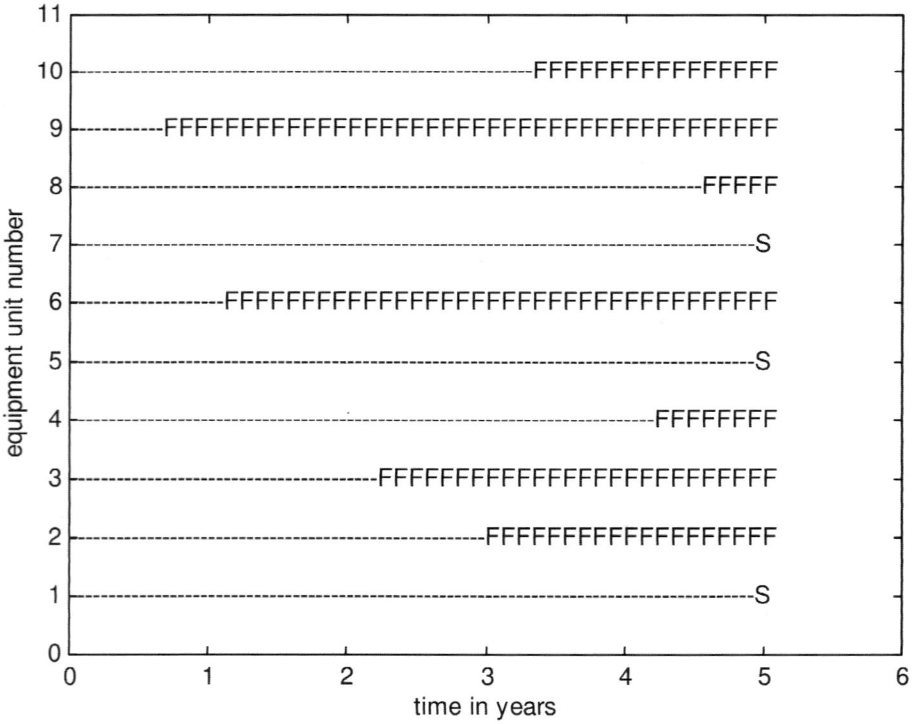

Figure 2. Example Time-to-Failures Spread Over 5 Years of Service

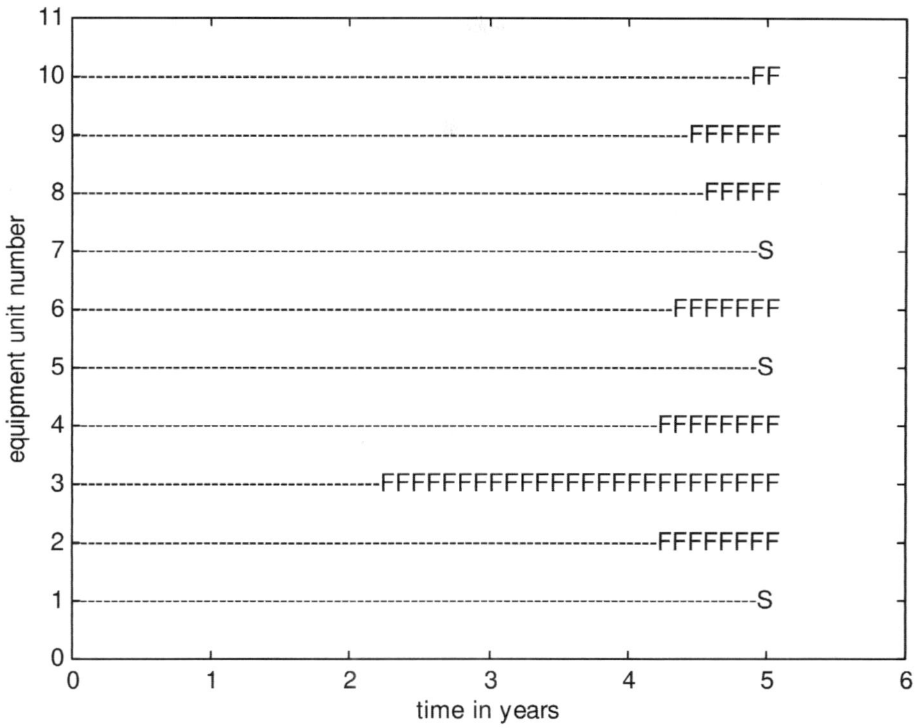

Figure 3. Example Time-to-Failures Clustered in Last Year of Service

period might exhibit one, two, or all three types of failure rates. The presence of all three types of failure rates gives rise to the traditional "bath tub curve" model for $\lambda_D(t)$.

Now consider the fact that the time-to-failure data captures all failures regardless of underlying cause. Therefore, $\lambda_D(t)$ estimated from the time-to-failure data along with PIF[2] is sufficient to model all types of failures. Given $\lambda_D(t)$ and PIF, and noting that R(0) = 1- PIF, PFDavg can be computed from (1) above for any valid $\lambda_D(t)$[3].

ESTIMATING $\lambda_D(t)$ FROM PROOF TEST DATA

Because the PRV is normally closed, the "stuck-shut" failure mode cannot be detected while the PRV is installed within the process being protected. Identifying this failure mode requires a proof test. In a proof test, the PRV is removed from the process and is pressurized to determine the pressure at which the PRV opens. If the test pressure

[2] For mathematical completeness, we state for the record that PIF could easily be incorporated in $\lambda_D(t)$ as an impulse (Dirac delta function) at t=0. Here we treat PIF separately for simplicity of explanation and because, as a practical matter, it would be handled separately in numerical computation of PFDavg.

[3] Reliability theory imposes certain constraints on $\lambda_D(t)$ which must be met. Specifically, $\lambda_D(t)$ must be ≥ 0 for all t and ${}_0\int^{\infty} \lambda_D(s)\, ds \rightarrow \infty$.

required to open the PRV the first time exceeds 150% of set pressure, the PRV is deemed to have "failed to open", i.e., the PRV is "stuck shut". This type of testing is costly and, while it identifies failed PRV, it does not provide an indication of *when* the failure occurred prior to the proof test.

Now consider the case where we do not have time-to-failure information but only proof test results. Re-examine Figures 1 - 3 and assume that we can only observe the state of each unit of equipment at t = 5 years. This is equivalent to conducting a proof test at T_P = 5 years and recording whether the unit successfully opened (S) or failed (F) to open. Note that the proof test results would be identical for all three distinctly different failure patterns. Thus, proof test data analysis, *based solely on the operating time accumulated prior to proof test and the outcome (S/F) of the test*, **cannot** identify a unique $\lambda_D(t)$ underlying the actual failure patterns since these patterns are not observable. However, subject to appropriate analysis, proof test data can produce a series of candidate $\lambda_D(t)$ and PIF combinations that are all statistically consistent with the observed proof test data. Each different candidate $\lambda_D(t)$ and PIF combination will potentially result in a different PFDavg. Determining which one of the candidate $\lambda_D(t)$ and PIF combinations best describes the true underlying failure pattern (and corresponding PFDavg) will require additional information and analysis techniques.

PROOF TEST DATA ANALYSIS METHODS

There are currently two analysis methods available for application to proof test data, Quantal Response Analysis (QRA) [2, 6] and Statistical Signature Analysis (SSA) [4, 5, 7]. The discussion below compares and contrasts them.

Quantal Response Analysis

In QRA, the analyst assumes the form $\lambda_D(t)$ will take. For example, $\lambda_D(t) = \lambda_D$, a constant, $\lambda_D(t) = At^n + b$, $n > 0$, a growth function, etc. The proof test data is then sorted and grouped, the grouped data is transformed, curve fitting methods are applied to find a best fit for the integral of $\lambda_D(t)$, and, if a valid curve fit is found, the unknown parameters in the assumed form of $\lambda_D(t)$ are estimated.

QRA has a number of drawbacks. It requires experience to appropriately group the data. The results of the analysis can be very sensitive to the particulars of the data grouping and different groupings may lead to different fitted curves. Because $\lambda_D(t)$ must fulfill certain conditions, the fitted curves are also constrained. As a consequence, it is sometimes difficult or impossible to find a curve fit that also meets the required constraints. When no valid curve fit is discovered, it does not mean that one does not exist; merely that one has not been found. Conversely, if only one valid curve fit is discovered, it does not mean it is the only valid fit; merely the only one that has been found. Further, when no valid curve fit is discovered, QRA does not indicate what to do next to try to find a valid fit. QRA requires that the proof test times, T_P, have a reasonable amount of variation in time before curve fitting can occur. For example, if an operating company faithfully carried out proof testing every 2 years plus or minus a month, the proof test times would be too tightly clustered to allow for effective curve fitting. Lastly, if new proof test data becomes available and is added to the data set, it is likely the data must be completely re-analyzed to take into account the new data.

Although in theory, QRA could handle any assumed valid form for $\lambda_D(t)$, as a practical matter, complex $\lambda_D(t)$ would lead to the need for very complicated curve fitting that would likely not be successfully achieved. To date, when applying QRA to proof test data, only two forms of $\lambda_D(t)$ have been considered individually: constant failure rate and failure rate growth. In particular, no attempt has been made to assume an infant mortality form for $\lambda_D(t)$ or to include more than one type of failure rate function in $\lambda_D(t)$, e.g., to include both a declining failure rate on one time interval and a constant failure rate over another time interval.

Statistical Signature Analysis

SSA takes a different philosophical approach than QRA to the task of analyzing proof test data. It is simply a "guess and check" method of exploring the proof test data. It allows the analyst to assume not only the form of $\lambda_D(t)$ but also to assume values for all of the parameters defining $\lambda_D(t)$ and to include either a zero or non-zero value for PIF. It then compares the number of failures actually observed in the proof test data to the number of failures that would likely be seen if the assumed $\lambda_D(t)$ and PIF were correct. Based on this comparison, a statistical signature is produced. The signature is valid if it fulfills certain properties in which case the assumed parameter values from $\lambda_D(t)$ and PIF are candidates to explain the observed data. If the signature is not valid, the assumed parameter values are not candidates to explain the observed data. However, the properties of the signature that make it invalid also indicate broadly what parameter values to try next in the search for a valid signature.

Unlike QRA, SSA does not require any data sorting. Even if the proof test times are highly clustered, SSA can process the data. The proof test data can be fed into the algorithm as it is collected and one point in the signature is computed and plotted for each proof test based on the results of the latest test and all previous tests in the data set. New data is simply added to the running analysis. There is no limitation on the complexity of $\lambda_D(t)$ that may be assumed and $\lambda_D(t)$ may be described mathematically piecewise over several intervals. Thus quite complex $\lambda_D(t)$ can be constructed from a series of simpler parts.

APPLYING SSA TO PREVIOUSLY ANALYZED PRV PROOF TEST DATA

Proof Test Data Analyzed

In [2, 3], three proof test data sets which met the data quality intent of the CCPS PERD Initiative[4] were analyzed using QRA to estimate the PRV constant failure rate, λ_D, and PIF. We now wish to revisit those data sets and apply SSA. Unfortunately, Data Set II from the original study is not retrievable in the form required for SSA. Thus, this paper is limited to re-analyzing Data Set I and used-valve[5] proof test data from a recently expanded Data Set III.

We have also gathered some additional general information about Data Sets I and III. Data Set I consists of PRV proof tests from a mix of clean and foul service, weighted toward clean service. These proof tests were primarily bench tests that included both in-house and out-sourced testing. The expanded Data Set III consists of PRV proof tests almost exclusively from clean service. All of the tests were in-house bench tests.

Assumed Form of $\lambda_D(t)$

SSA is capable of incorporating $\lambda_D(t)$ of any complexity and this paper reports on the application of SSA using a more complex form of $\lambda_D(t)$ than has previously been studied. Although we originally intended to include all characteristics of the traditional "bath tub curve" in our $\lambda_D(t)$, a closer examination of the two data sets to be utilized in the analysis showed that neither set contained any evidence of failure rate growth indicating end of life failures. Consequently, we limited the characteristics of $\lambda_D(t)$ in this study to a declining failure rate (modeling infant mortality over the time interval $[0, T_{IM}]$) in combination with a constant failure rate. Two different declining failure rates were considered, exponential and linear, to represent steep or shallow rates of decline. Clearly, other variations are possible. The two different forms of $\lambda_D(t)$ studied are illustrated in Figures 4 and 5 and their formulas with relevant parameters are given in Equations (4) and (5). For exponentially declining infant mortality

$$\begin{aligned} \lambda_D(t) &= c\, e^{-kt} && \text{for} \quad 0 \le t \le T_{IM} \\ &= \lambda_D && \text{for} \quad T_{IM} \le t \le \infty \end{aligned} \qquad (4)$$

For linearly declining infant mortality

$$\begin{aligned} \lambda_D(t) &= d + m\, t && \text{for} \quad 0 \le t \le T_{IM} \\ &= \lambda_D && \text{for} \quad T_{IM} \le t \le \infty \end{aligned} \qquad (5)$$

[4] See Acknowledgement for contact information for and additional description of the CCPS PERD Initiative.

[5] Data set III contained test data from both new and used valves. We considered only the used-valve data as the new valves had no operating time on them.

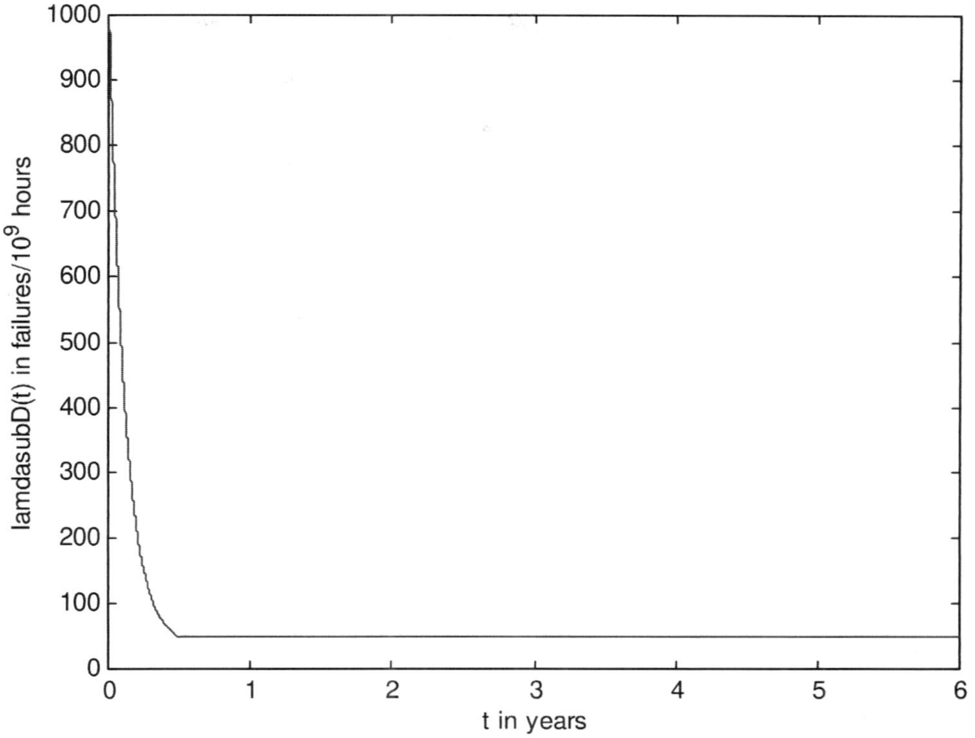

Figure 4. Exponentially Declining Infant Mortality and Constant Failure Rate

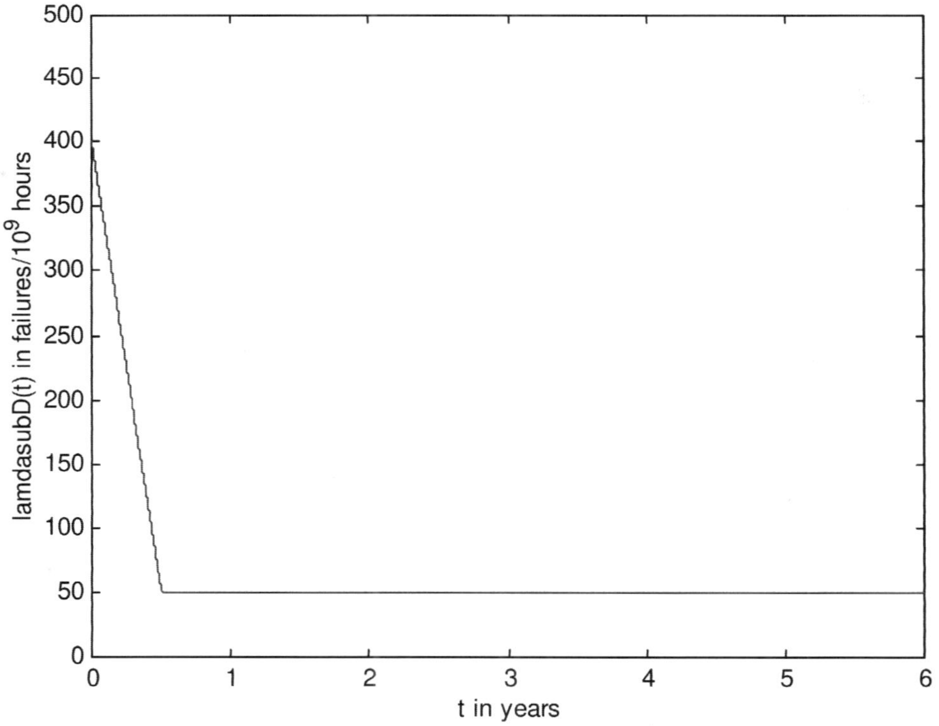

Figure 5. Linearly Declining Infant Mortality and Constant Failure Rate

It appears that in either case, when an analyst applies SSA to the proof test data s/he must provide the values for 5 parameters: PIF, λ_D, T_{IM}, and c and k, or d and m. However, this is reduced to 4 values because of the requirement of continuity at the time value T_{IM}.

From (4)

$$\lambda_D(T_{IM}) = c\, e^{-kTIM} = \lambda_D \qquad (6)$$

which implies that

$$k = - (1/T_{IM}) \ln (\lambda_D / c) \qquad (7)$$

and from (5),

$$\lambda_D(T_{IM}) = d + m\, T_{IM} = \lambda_D \qquad (8)$$

which implies that

$$m = (\lambda_D - d)/T_{IM}. \qquad (9)$$

Choosing Parameter Values

Equipment failures due to the constant failure rate, λ_D, are usually considered to be caused by environmental stresses that exceed the average strength of the equipment during its normal lifetime. As such, we would expect that λ_D would be ever-present. Thus, in this study, λ_D always has a non-zero value. However, all other unconstrained parameters may assume either zero or non-zero values. In order to limit the number of parameters to be varied, we decided to fix T_{IM}, the end of the period of infant mortality failures at 0.5 years for this study. Note that only one infant mortality model can be considered at a time. Consequently, c and d cannot simultaneously assume non-zero values. Table 1 shows the combinations of parameter values considered.

Table 1. Combinations of Parameter Values Considered

λ_D	PIF	c	d
non-zero	0	0	0
non-zero	non-zero	0	0
non-zero	0	non-zero	0
non-zero	0	0	non-zero
non-zero	non-zero	non-zero	0
non-zero	non-zero	0	non-zero

Methodology and Results

To conduct this study we began with Data Set I and sought a value of λ_D that alone could account for all the failures observed in the proof test data, i.e., SSA was run with all parameter values other than λ_D set equal to zero. Figure 6 shows the statistical signature produced with λ_D = 514 FITS. We then selected other smaller values for λ_D and, by trial and error, found various combinations of other parameter values that resulted in statistical signatures similar to the one in Figure 6. Since these combinations of values produced similar signatures, they were deemed to be equally good explanations for the observed data.

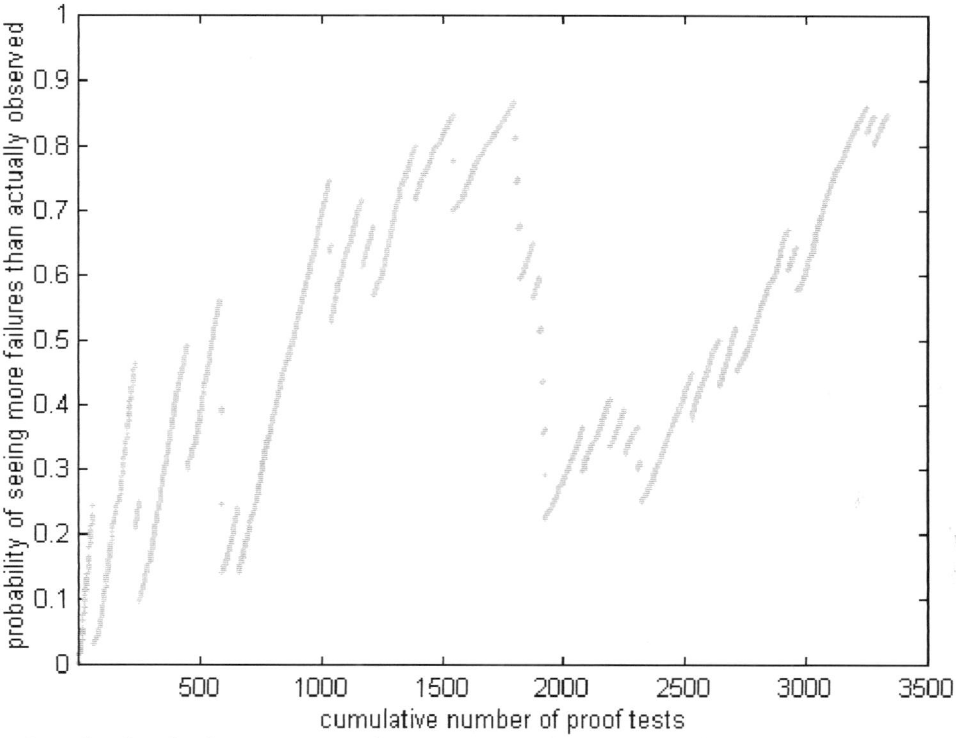

Figure 6. Statistical Signature: λ_D = 514 FITS and All Other Parameters Zero

We studied a wide variety of λ_D values including 400, 300, 200, 100, 75, 50, and 25 FITS producing far too much data to include in this paper. We illustrate our findings for λ_D = 50 FITS, chosen because it is the midrange value for the generally accepted range of λ_D and this range is supported both by previous QRA and by FMEDA analyses [8]. Table 2 shows a sampling of combinations of parameter values. Note that these are only a sampling and many other combinations were discovered for any fixed value of λ_D. Figure 7 shows the signatures from all combinations in Table 2 overlaid, with the signature for Case A plotted in green and the signatures for Cases B - F plotted in blue. Note that the signatures for Cases B - F are virtually indistinguishable from one another

Table 2. Sampling of Parameter Values Offering Equally Good Explanations of the Proof Test Data in Data Set I

Case	λ_D FITS	PIF	c FITS	d FITS
A	514	0	0	0
B	50	0.010065	0	0
C	50	0	13,500	0
D	50	0	0	4795
E	50	0.002	10,400	0
F	50	0.002	0	3860

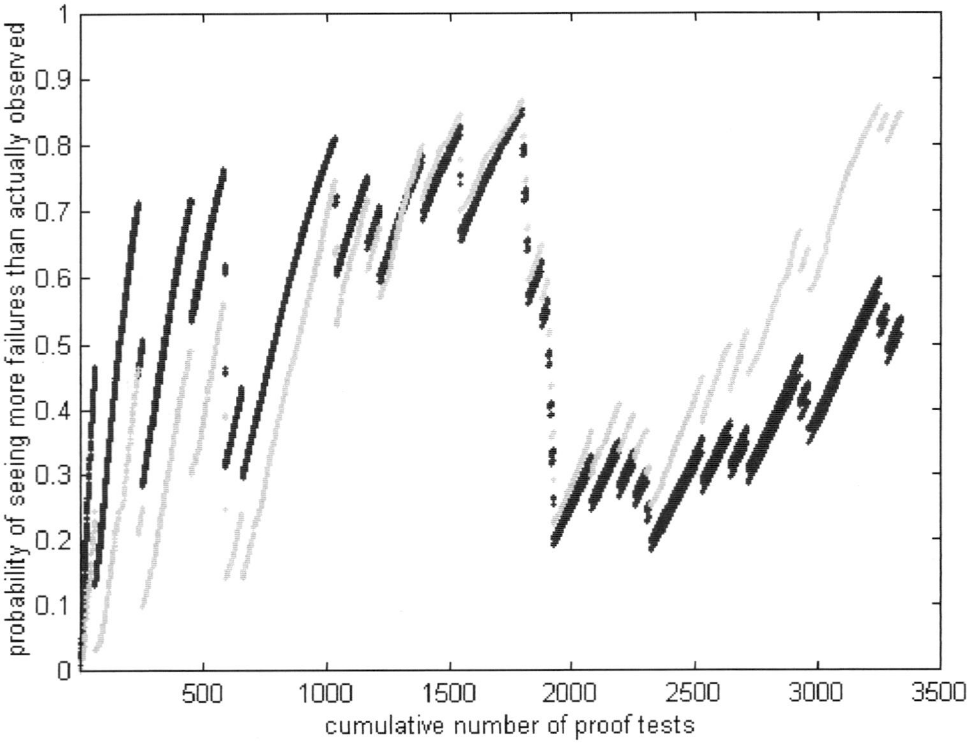

Figure 7. Overlay of All Statistical Signatures for Combinations in Table 2

and are very similar to the signature for Case A. (If Figure 7 is viewed in black and white, Case A is the lighter, narrower line that deviates from the darker, thicker line.)

We preceded similarly with the used-valve test data from the extended Data Set III. Table 3 shows a sampling of the combinations that produced essentially the same signature and Figure 8 shows all of the signatures overlaid. Note that in this case, the signature with λ_D = 500 and other parameters set to zero (Case G) is shown in green while the other cases from Table 3 are shown in blue. (If Figure 8 is viewed in black and white, Case G is the lighter line which deviates from the darker, thicker lines.)

Table 3. Sampling of Parameter Values Offering Equally Good Explanations of the Proof Test Data in Data Set III

Case	λ_D FITS	PIF	c FITS	d FITS
G	500	0	0	0
H	50	0.01289	0	0
I	50	0	19,075	0
J	50	0	0	6,450
K	50	0.005	10,720	0
L	50	0.005	0	3,975

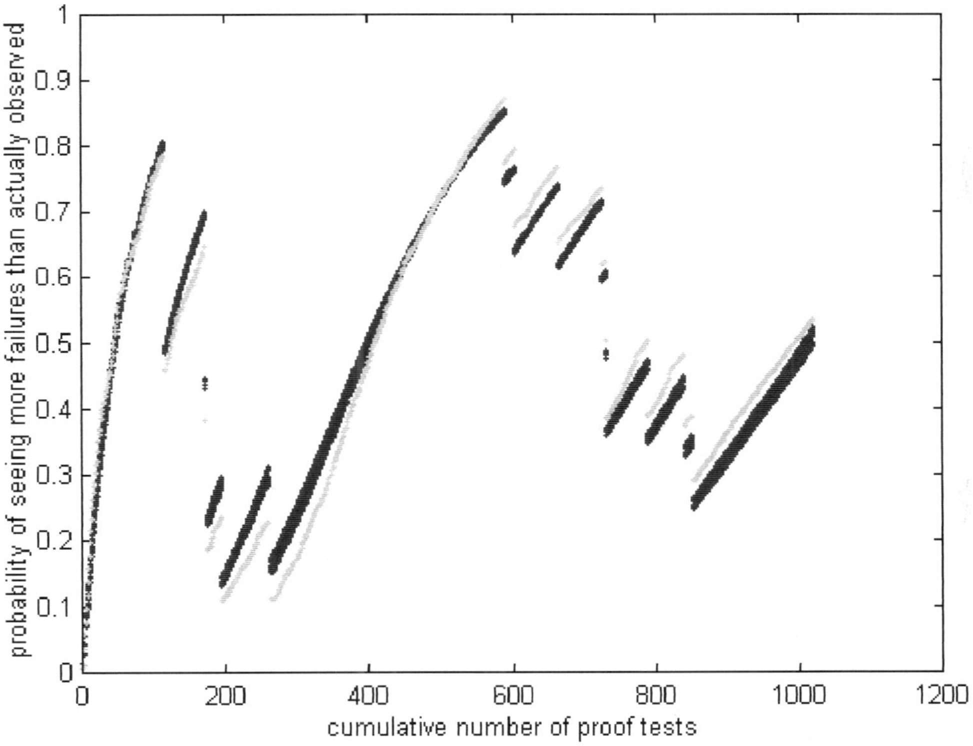

Figure 8. Overlay of All Statistical Signatures for Combinations in Table 3

IMPLICATIONS AND RESULTS FOR PFDavg

PFDavg is correctly defined in (1) for any valid $\lambda_D(t)$. However, with more complicated $\lambda_D(t)$, the actual calculation of PFDavg for any value of T_P becomes more complex and may require numerical integration of all or part of the integral. The exact formulas for PFDavg are derived in [7] for the cases of $\lambda_D(t)$ given in (4) and (5) and are the means by which we computed PFDavg for Figures 9 and 10 below.

Given that each of the combinations of parameter values in Table 2 or in Table 3 produces essentially the same statistical signatures, all are equally good explanations for their respective observed failure data. But each combination of parameter values potentially produces a different assessment of PFDavg. Figures 9 and 10 show plots of PFDavg vs. T_P for the parameter combinations given in Tables 2 and 3, respectively.

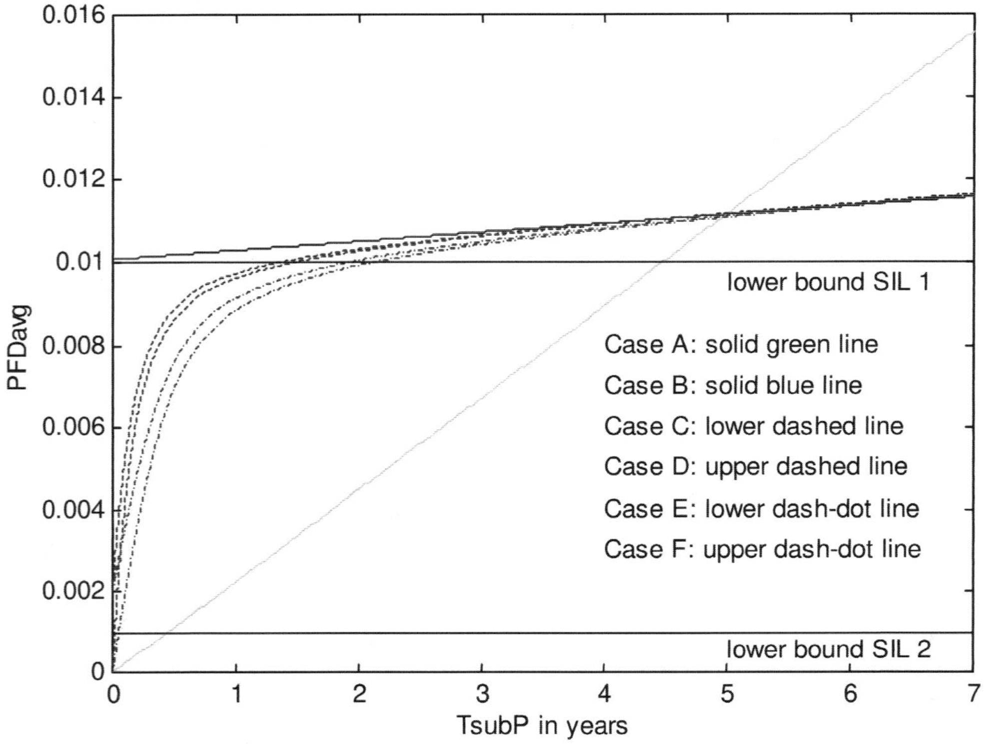

Figure 9. Plots of PFDavg for All Combinations in Table 2

In Figure 9, note that for Cases B – F, which all have λ_D = 50 FITS, the PFDavg plots are discernibly different for values of T_P < 1 year, but show little difference for T_P between 2 and 4 years. Once T_P > 4 years, the PFDavg computed for these cases are approximately equal. For T_P < 1 year, Cases C – F support SIL 2 behavior. For T_P > 4.5 years, the data support SIL 1 behavior only regardless of the model used to explain the observed failures in Data Set I.

Figure 10 shows that Cases I – L support SIL 2 behavior only for values of T_P < 6 months. For T_P > 6 months, the data support SIL 1 behavior only regardless of the model used to explain the observed failures in used-valve data from extended Data Set III.

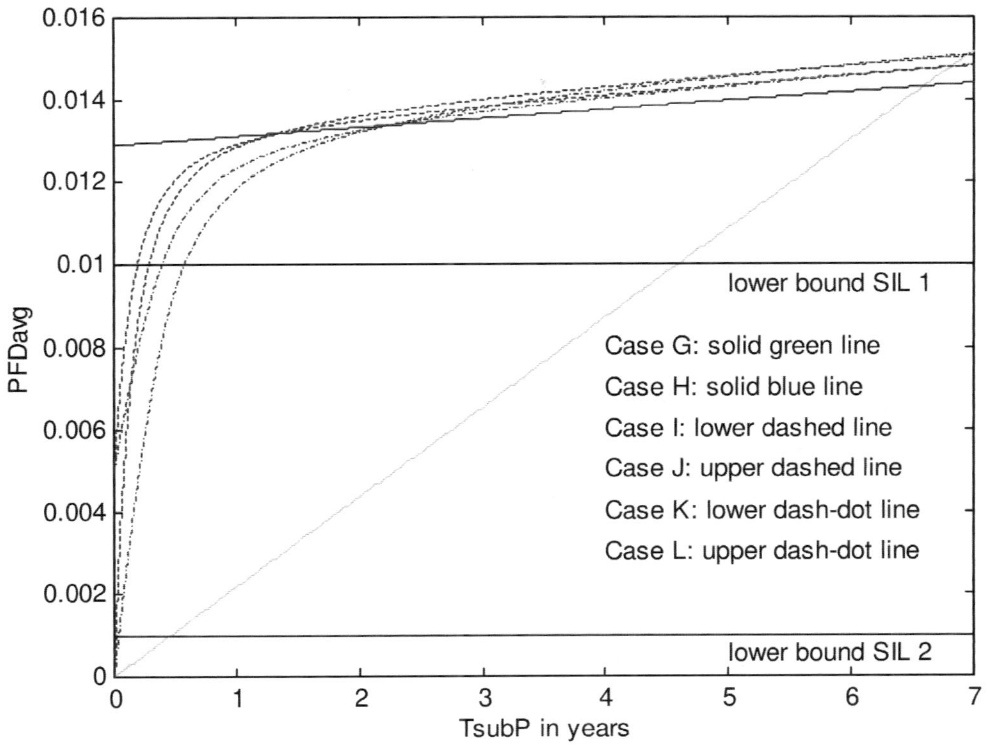

Figure 10. Plots of PFDavg for All Combinations in Table 3

DISCUSSION

Clearly, the SSA method opens up many new possibilities for modeling equipment failure rates from proof test data. Its current shortcoming is that, alone, it cannot tell us which one of the many models, identified as being statistically consistent with the data, actually describes the true underlying failure mechanisms generating the observed proof test data. It is likely that other independent means to estimate or predict some of the parameters in the $\lambda_D(t)$ model will narrow the possibilities that need to be considered. For example, by using FMEDA techniques to predict the constant portion of the failure rate, the remaining parameter ranges may be significantly reduced. Investing in the development of additional analysis techniques is worthwhile as these techniques will likely produce useful models that allow designers to better optimize safety system designs. Finally, finding ways to identify the $\lambda_D(t)$ which most accurately explains the underlying failure mechanisms is also important when we seek to reduce the occurrences of dangerous failures.

The results reported here apply to the particular data sets analyzed. We do not know what results or conclusions might be observed if additional data sets become available for analysis. But it is critical that *appropriate* data be collected *and* analyzed and the results shared. There is significant potential to reduce costs while maintaining safety levels or even reduce costs while improving safety if we can accurately understand and identify the underlying sources of dangerous failure.

REFERENCES

1. "TUV Certificate" (2006), Report No. 432/21205342, 30.01.2006, TUV Rheinland Group.

2. Bukowski, J. V. (2007), "Results of Statistical Analysis of Pressure Relief Valve Proof Test Data Designed to Validate a Mechanical Parts Failure Database," Technical Report, September, exida, Sellersville, PA.

3. Bukowski, J. V., and Goble, W. M. (2009), "Analysis of Pressure Relief Valve Proof Test Data," *AIChE Journal Process Safety Progress*, March 2009.

4. Bukowski, J. V. (2008), "A New Method for Analyzing Proof Test Data," Technical Report, September, exida, Sellersville, PA.

5. Bukowski, J. V. and Goble, W. M. (2009), "Unexpected Results from the Analysis of PERD Test Data & the Implication for Pressure Relief Valve Safety," *Proceeding 64th Annual Instrumentation Symposium for the Process Industries*, January, College Station, Tx.

6. Sheesley, J. H., Thomas, H. W. and Valenzuela, C. A. (1995), "Quantal Response Analysis of Relief Valve Test Data," *ASQC 49th Annual Quality Congress Proceedings*, Cincinnati, OH, pp. 741-748.

7. Bukowski, J. V. (2008), "Statistical Signature Analysis: Modeling Complex $\lambda_D(t)$ from Proof Test Data and the Effects on Computing PFDavg," Technical Report, November, exida, Sellersville, PA.

8. Bukowski, J. V. and Goble, W. M. (2009), "Validation of a Mechanical Component Constant Failure Rate Database," *Proceedings Annual Reliability and Maintainability Symposium*, January, Fort Worth, TX.

ACKNOWLEDGEMENTS

The authors wish to thank the CCPS PERD Initiative for providing PRV proof test data used in the data analysis described in this paper. This initiative is comprised of operating companies, consultants, equipment manufacturers, and engineering companies committed to developing the technical guidance and tools necessary to capture on-going performance data, enabling operation of an equipment reliability database. The objective of PERD is to make high quality, valid, and useful data available to participating HPI and CPI companies to support availability, reliability, and equipment design improvements, maintenance strategies, and life cycle cost determination. For more information, contact Dave Belonger, CCPS PERD staff consultant, at dbelonger@verizon.net, or Hal Thomas, PERD Chairman, at thomashw@airproducts.com.

The authors also wish to thank Robert Gross of Savannah River Nuclear Solutions for providing additional proof test data analyzed in this paper and for his helpful discussions and insights into proof test practices.

Validation of Spring Operated Pressure Relief Valve Time-to-Failure and the Importance of Statistically Supported Maintenance Intervals

ABSTRACT

The Savannah River Site operates a Relief Valve Repair Shop certified by the National Board of Pressure Vessel Inspectors to NB-23, The National Board Inspection Code. Local maintenance forces perform inspection, testing, and repair of approximately 1200 spring-operated relief valves (SORV) each year as the valves are cycled in from the field.

The Site now has over 7000 certified test records in the Computerized Maintenance Management System (CMMS); a summary of that data is presented in this paper. In previous papers, several statistical techniques were used to investigate failure on demand and failure rates including a quantal response method for predicting the probability of failure in the interval {0, t} time in service. The non-conservative failure mode for SORV is commonly termed "stuck shut"; industry defined as the valve opening at greater than or equal to 1.5 times the cold set pressure. Actual time to failure is typically not known, only that failure occurred some time since the last proof test (censored data).

The resulting Monte Carlo Simulation (MCS) in this paper shows that previous conclusions on the statistical lifetime predictions and failure studies were not as conservative as initially thought. The MCS employs an aging model for lift pressure increase as a function of set pressure, valve manufacturer, and time. This paper attempts to answer two questions: "what is the predicted failure rate over the chosen maintenance/ inspection interval", and "do we understand aging and variability sufficiently to estimate risk when basing proof test intervals on proof test results."

Stephen P. Harris, Ph.D.

Savannah River National Laboratory
Aiken, South Carolina
stephen.harris@srnl.doe.gov

Robert E. Gross, P.E.

Savannah River Nuclear Solutions
Aiken, South Carolina
robert.gross@srs.gov

SUMMARY

The ratio (*R*) of relief valve Test Pressure to Set Pressure (TP/SP) greater than or equal to 1.5 in the as-found condition is considered by industry and API 576 to be "stuck shut." When a relief valve is removed from service for maintenance, the as-found lift pressure or "proof test" result in general correlates with the valve's performance "on demand." During an actual over pressure event failing to open by 1.5 times the set pressure will challenge process piping and vessel integrity. The Monte Carlo simulation (MCS) in this report was based on a statistical model developed from new valve lift pressure (proof test) results. These include new valves shipped from the manufacturer and those that were rebuilt, i.e., returned to "like new" condition after having been in service. Hypothetical yearly percentage increases (1%-2%) in lift pressure were built into the statistical model for testing the analysis methods.

INTRODUCTION

The Savannah River Site (SRS) joined the Process Equipment Reliability Database (PERD) Committee 18 on pressure relief valves in 2005. Members who contribute test data in a PERD approved format are able to access available industry wide data. SRS first uploaded 4000 data points in early 2007. Consequently, PERD shared several data sets they had accumulated since 1993 with SRS. At the time, there were 9 active members of the Center for Chemical Process Safety (CCPS) within PERD.

PREVIOUS WORK - SRS and PERD

The PERD data set provided an estimated Probability of Failure on Demand (PFD) of approximately 1.2 % while the estimated value for SRS data was approximately 0.7 %, see Table 1. A comparative representation of the ratio distribution is shown in Plot 1. Evaluation of differences causing the spike in SRS ratios between 1.10 and 1.20 was necessary. A review of SRS data by manufacturer determined that small, inexpensive, soft-seated style valves represented 65% of the valves in the 1.10 < *R* < 1.20 range. PERD data provided similar average R for soft seat valves but the number tested is **<** **½** % while soft seat for SRS comprise more than 7% in the tested population.

Table 1 Overall Proportion of Valves that Failed up to 5 years in Service (n is the number of proof tests with R≥1.5)

Source	N	n	p (%)	95% Confidence Limits
SRS	2472	17	0.69	[0.40%, 1.10%]
PERD	6768	84	1.24	[0.99%, 1.53%]

Table 2. Summary of Proof Test Statistics for PERD vs. SRS All Proof Test Data Combined, New and Used

Source	Number of Valves	Mean	Standard Deviation
SRS	2472	1.034	0.090
PERD	6768	1.006	0.124

Box and Whisker plots based on the ratio *R* for PERD and SRS data are shown in Plot 2 and the means and standard deviations are listed in Table 2. PERD recorded a significant amount of valves classified as stuck shut (*R*≥1.5); almost twice the rate as SRS, but their mean value is still 2% lower. SRS data indicated that the 95% confidence interval for probability of failure on demand (PFD) is 0.40% to 1.10% while the PFD for PERD is 0.99% to 1.53%. Overlap of the intervals implies that the proportion of valves that failed by source is equivalent. In Plot 2, note the close groupings of test data at approximately R=1.00; 31% of PERD Data has *R*=1.00 vs. 19% for the SRS Data. In the SRS data, we observed a failure rate of 1.3% /Year from 1 to 5 years time in service (Plot 3).

RISK ASSESSMENT

How does estimated or perceived SORV performance compare to actual proof test data results? It has been stated that an intuitive or subjective estimate of the PFD for a standby system *might double* if the standby time doubles. [1] If the PFD were actually shown to double with twice the time in service by the actual test data, then would not the estimated process safety risk increase significantly? Analysis of test data available through the CCPS-PERD, including SRS to date does not establish a statistically significant increase in the failure rate from 1 to 5 years in service. Several different authors in different publications have concluded the failure rate is fairly constant over the period.[2, 3, 4, 5] How would a Monte Carlo simulation (MCS) of the representative data using an aging model *substantiate the failure rate over time - beyond 5 years?*

This paper investigates the consequences of standard statistical methods applied to PSV data using simulation. It explores the use of regression analysis and a quantal response method when applied to simulated data.

DATA SUMMARY

The 9 manufacturers representing about 85% of the SRS valve populations are displayed in Table 3. KFC represents the manufacturer with the most valves (nearly 20%) in the new valve population while representing only 9% of the used valve population. The top manufacturer in the used valve population is DVL representing nearly 17 % of the population. The differences in the new and used valve distributions across the different manufacturers are accounted for in the MCS.

New valves are tested before installation at SRS. A valve is released for installation at SRS if the actual lift pressure (proof test) is within $\pm$3% of the set pressure. If a test is outside of this range, the valve is adjusted and then retested.

Table 3 - Number and Percentage (%) of New and Used Valve Population for Major Manufactures

Manufacturer	N (new)	% (new)	N (used)	% (used)
CID	144	6.2%	70	7.8%
CS	330	14.1%	140	15.5%
CD	298	12.8%	54	6.0%
CRL	139	6.0%	88	9.8%
DVL	162	6.9%	152	16.9%
KFC	464	19.9%	79	8.8%
KVD	204	8.7%	67	7.4%
TNF	118	5.1%	123	13.6%
WRT	118	5.1%	0	0.0%
Total	1977	84.7%	773	85.7%

FAILURE TIMES – SURVIVAL FUNCTIONS

A model is needed for the MCS to generate ratios that are realistic, include uncertainty and also the impact of specified trends. The SRS new valve data base was used for this purpose. The resulting model represents lift pressure (psig) as a function of set pressure and manufacturer. The model is expressed in the natural logarithmic space as:

$$\log_e(y_{Test\,Pressure}) = \beta_0 + \beta_1 \log_e(y_{Set\,Pressure}) + Manufacturer\ Terms + \varepsilon \qquad (1)$$

where $\log_e(y)$ is the natural logarithm of y, $y_{Test\,Pressure}$ is the Test Pressure (psig), $y_{Set\,Pressure}$ is the set pressure (psig) and ε is the random error assumed to follow a normal distribution with zero mean and constant standard deviation σ. Using Model (1), $y_{Test\,Pressure}$ can be represented in the original metric (psig) as:

$$y_{Test\,Pressure} = e^{[\beta_0+\beta_1 \log_e(y_{Set\,Pressure})+Manufacturer\ Terms+\varepsilon]} = e^{(\beta_0+Manufacturer\ Terms)}e^{\varepsilon}y_{Set\,Pressure}^{\beta_1} \quad (2).$$

The parameters in Model (1) were estimated using the JMP statistical software (SAS Institute, Cary, NC) in conjunction with the new valve data base. The estimates and standard errors (Plot 3) are:

$\hat{\beta}_0 = b_0 = 0.108 \pm 0.011$, $\hat{\beta}_1 = b_1 = 0.988 \pm 0.002$ and $\hat{\sigma}^2 = 0.070$.

The correlation between b_0 and b_1 is $Corr(b_0, b_1) = -0.84$ and is used in the MCS. The *Manufacturer Terms* essentially offset the intercept term β_0 in Model 1. Their estimates were included in the MCS but were not included in this report because of space requirements with there being 45 manufacturers. Model (2) served as the basis for the MCS after known trends were built into the model. The resulting model was placed in the used valve data base so that the distribution of manufacturers in service at the SRS could be represented.

MONTE CARLO SIMULATION (MCS)

All valves taken out of service for preventative maintenance have their service time recorded in the used valve data base. The exact impact of aging on valve performance within the PM interval (t_0, t_{PM}) where $t_0 = 0$ and t_{PM} is not known. For purposes of the MCS, it is assumed that the lift pressure could have been attained at any time t between installation and the PM maintenance time (time in service) $0 < t \le t_{PM}$. There is essentially no further aging impact in the MCS after time t and before the PM maintenance at time t_{PM}.

The time t is not known for valves in service. However, it is specified in the MCS. The aging model was taken to be linear with two cases simulated: $p = 1\%$/year and $p = 2\%$/year where p is the percentage increase in lift pressure per year. The aging model can be represented as:

$$y_{Test\,Pressure} = e^{(\beta_0+Manufacturer\ Terms)}e^{\varepsilon}y_{Set\,Pressure}^{\beta_1} + p \times t \times y_{Set\,Pressure} \quad (3)$$

MCS time was taken from a uniform distribution over the interval $0 < t \le t_{PM}$, while all other terms in Model 3 were generated from a normal distribution with means, standard errors and $\hat{\sigma}^2$ taken from the regression output (Plot 4).

REGRESSION ANALYSIS

The MCS was based on 25 runs for each p. The regression model $R = \beta_0^{'} + \beta_1^{'}t + \varepsilon^{'}$ where the ratio $R = y_{Test\ Pressure} / y_{Set\ Pressure}$ was fit to MCS time (t) using the JMP software. Time was first taken to be the time in service and then taken to be time t at which the ratio is attained over $0 < t \leq t_{PM}$. An illustration of one MCS run (of the 25) is displayed in Plot 5a for time t over $0 < t \leq t_{PM}$ and in for service time in Plot 5b. The aging rate was estimated as 1.13% / year increase (Plot 5a) and as 0.58%/ year (Plot 5b). The true value was 1%/year in the MCS. Over the 25 run MCS, the average increase in *R* is 0.56%/ year (Table 4) when using Service time as the regressor with a between run standard deviation of 0.0012. Using simulated MCS time at which *R* is incurred yielded an estimate of 1.08%/ year with a between run standard deviation of 0.0014. The true MCS value was a rate of 1% /year. Next, the true MCS value was taken to be a 2% /year increase. Over the next 25 runs, the average increase in *R* is 1.06%/ year when using the Service time as the regressor with a between run standard deviation of 0.0016. Using the simulated time at which *R* is incurred yielded an estimate of 2.06%/ year with a between standard deviation of 0.0015.

Table 4. Final Simulation Results Based on 25 Replications

Simulated % Increase/ Year	Estimated % Increase/ Year Using The Service Time	Estimated % Increase/ Year Using MCS Time
1%	0.56%	1.08%
2%	1.06%	2.06%

QUANTAL RESPONSE ANALYSIS

A PRV is considered as failed for the purposes of this MCS when $R \geq 1.10$ to ensure an adequate number of valves for statistical analysis. During pressure testing, if a valve has failed, one only knows that its failure time was before its maintenance inspection time. Similarly, if a valve proof test was acceptable, then its failure time is beyond the maintenance inspection time. Such test data are sometimes called quantal response (QR) data. [6] Data of this type are sometimes incorrectly analyzed by taking the inspection time to be the failure time. Illustrations of PRV quantal response analysis are described by and Sheesley[2] et. al., and Bukowski. [3]

To implement a QR analysis, the data are grouped into *m* consecutive non-overlapping intervals $I_1, I_2, I_3, \ldots I_m$ fully covering the distribution of maintenance times and $I_i = \{t; t_i \le t < t_{i+1}\}$ for $i = 1,2,3,...,m-1$. In practice, each interval should contain at least 5 failures; their occurrences are well represented in the MCS. The fraction failed, q_i, and the average time, T_i, for the failed valves are calculated for each interval. Once calculated, a plot is made of the points $-\log_e(1-q_i)$ versus T_i. End of useful life could be indicated by the time when the slope substantially increases after a period of constant failure rate.[5]

The MCS was driven by Equation 3 but with time $t \ge 2$ years. Therefore the QR plot should be flat over the first 2 years and then experience a positive slope over the subsequent years. The resulting QR plots are displayed in Plot 6a and 6b for a 1%/year increase and in Plot 7a and 7b for a 2%/year increase. A plot of q_i vs. time in service (Plot 6a) does not show the full impact of the increase in the ratio as when the MCS times are used (Plot 6b). One could come to the same conclusion for a 2%/year increase as displayed in Plot 7a vs. 7b. The change in slopes in the QR function is much more obvious when the MCS times are used (Plot 6b and 7b).

CONCLUSIONS

The primary result of the MCS was that an increased rate in the ratio (*R*) can be underestimated by a factor of approximately 2 if time in service is used in a regression analysis of the ratio (R) over time. Furthermore, quantal response (QR) analysis underestimates the impact of a 1% -2% / year increase in the ratio and is insensitive to the time of occurrence. Actual proof test data obtained by increasing maintenance intervals would provide additional insight into valve aging and end of useful life. At present there appears to be an abundance of proof test results and failure data out to 4 - 5 years, but few qualified data points to validate statistical simulation of the 6 - 10 year service life.

REFERENCES

1. Urbanik, Stanley A. (2004), "Evaluating Relief Valve Reliability When Extending the Test and Maintenance Interval", *Process Safety Progress (Vol.23, No.3),* September 2004 pp 191-196.

2. Sheesley, J.H, Thomas, H.W. and Valenzuela, C.A. (1995) "Quantal Response Analysis of Relief Valve Test Data", *ASQ 49th Annual Quality Congress Proceedings,* pp741-748

3. Bukowski, J. V. (2007), "Results of Statistical Analysis of Pressure Relief Valve Proof Test Data Designed to Validate a Mechanical Parts Failure Database," *Exida Technical Report,* September, Sellersville, PA.

4. Gross, Robert E., Harris, Stephen P, (2006) "Extending Pressure Relief Valve Inspection Intervals by Using Statistical Analysis of Proof Test Data", *Proceedings of the ASME PVP-2006 / ICPVT – 11 Conference,* Vancouver, BC, Canada July 23-27, 2006.

5. Bukowski, J. V. and Goble, W. M., "Statistical Analysis of Pressure Relief Valve Proof Test Data: Findings and Implications," *Proceedings AIChE 4th Annual Global Conf on Process Safety,* New Orleans, LA, April, 2008.

6. Nelson, W.B. (1982), "Applied Life Data Analysis", New York, John Wiley & Sons.

Plot 1: Pressure Relief Valve Proof Test Results at TP/SP Ratios > 1.10
Percentage of valves tested vs. Proof Test Ratios for SRS and PERD Data

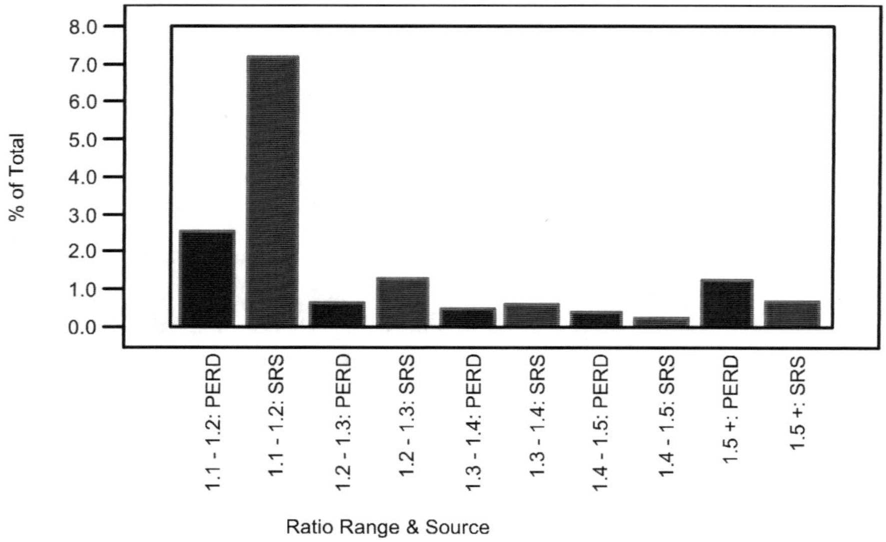

Plot 2 - PERD vs. SRS PSV Data New & Used PSVs Combined
Test Pressure/Set Pressure (R) by Company

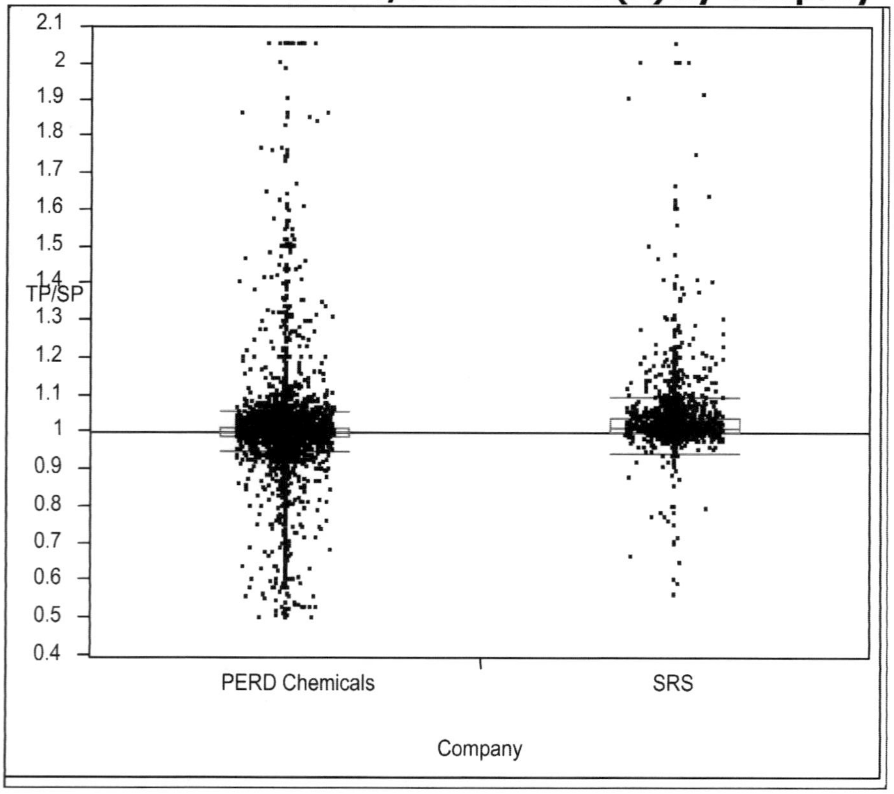

Plot 3

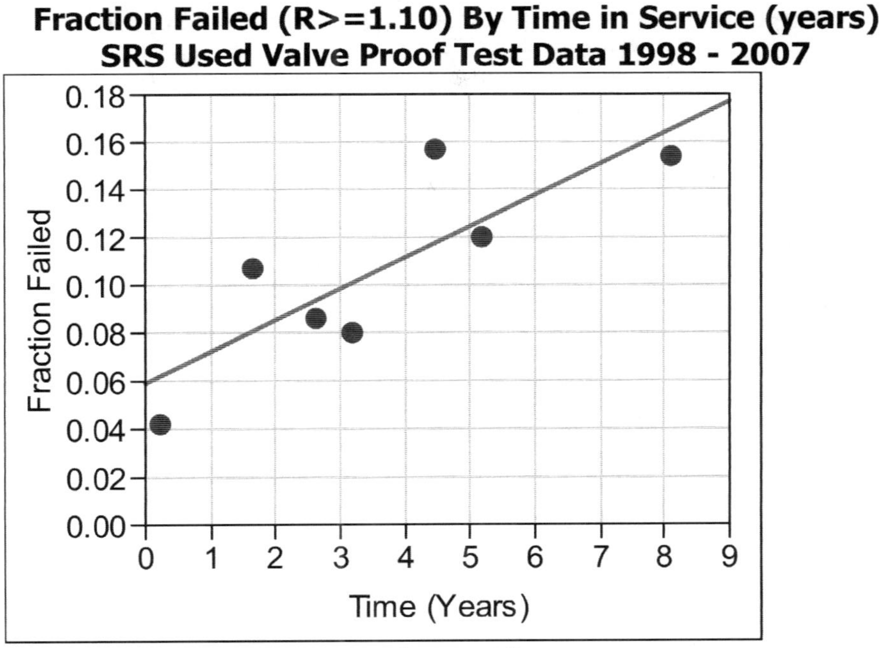

Linear Fit

Fraction Failed = 0.059 + 0.0132*Time (Years)

Failure Rate=1.3%/year

Summary of Fit

RSquare	0.679
Root Mean Square Error	0.0256
Mean of Response	0.1065
Observations (or Sum Wgts)	7

Analysis of Variance

Source	**DF**	**Sum Squares**	**of Mean Square**	**F Ratio**
Model	1	0.0069	0.00694	10.6
Error	5	0.0033	0.00066	**Prob > F**
C. Total	6	0.0102		0.023

Parameter Estimates

Term	**Estimate**	**Std Error**	**t Ratio**	**Prob>|t|**
Intercept	0.059	0.0176	3.33	0.021
Time (years)	0.0132	0.0041	3.25	0.023

Plot 4
Regression Model based on New Valve Data Set
Predicted Lift Pressure By Actual Lift Pressure

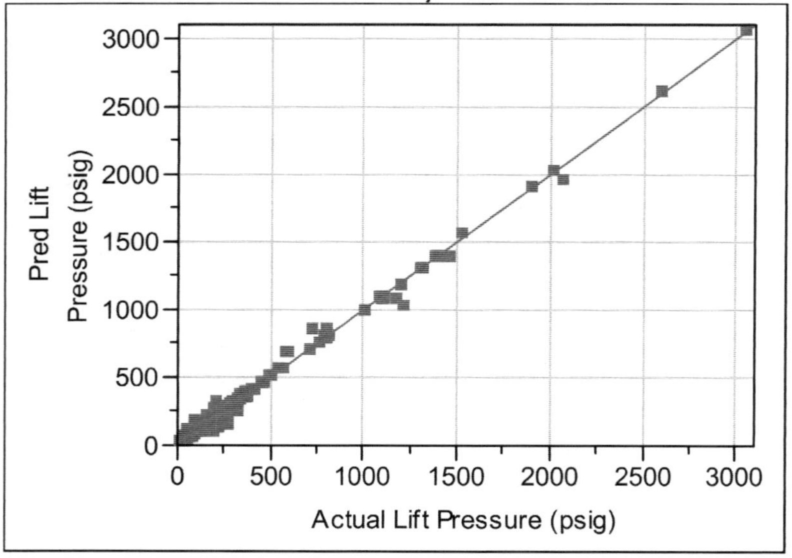

Response log$_e$(Lift Pressure)
Summary of Fit

RSquare	0.99
Root Mean Square Error	0.07
Mean of Response	4.64
Observations	2338

Analysis of Variance

Source	DF	Sum of Squares	Mean Square	F Ratio
Model	45	1483.8	32.97	6722.4
Error	2292	11.2	0.005	Prob > F
C. Total	2337	1495.1		0.0000

Parameter Estimates

Term	Estimate	Std Error	t Ratio	Prob>|t|
Intercept	0.108	0.0108	10.0	<.0001
log$_e$(Set Pressure)	0.988	0.0019	508.8	0.0000
+ Manufacturer Terms				

Effect Tests

Source	Nparm	DF	Sum of Squares	F Ratio	Prob > F
In Set	1	1	1270.0	258921.3	0.0000
Manufacturer	44	44	1.3	5.7924	<.0001

Plot 5a

Regression Estimates using One Iteration of the Simulation with 1%/ Year Deterioration Using MCS Time (Years)

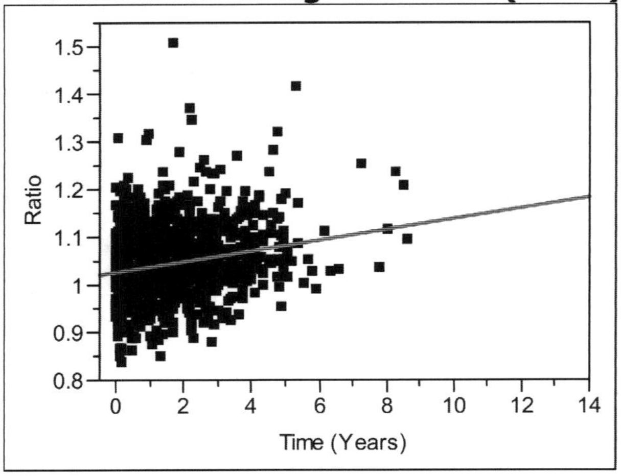

Ratio = 1.027 + 0.0113*Time (Years)

Parameter Estimates

Term	**Estimate**	**Std Error**	**t Ratio**	**Prob>|t|**
Intercept	1.027	0.0042	245.8	0.0000
Time (Years)	0.0113	0.0018	6.2	<.0001

Plot 5b

Regression Estimates using One Iteration of the Simulation with 1%/ Year Deterioration using Time In-Service (Years).

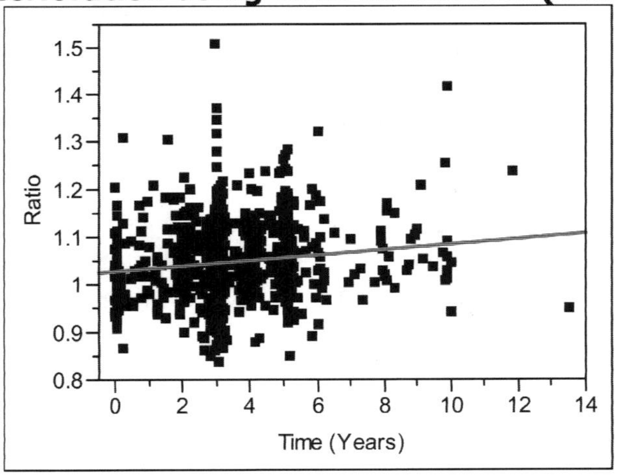

Linear Fit

Ratio = 1.026 + 0.0058*Time (Years)

Parameter Estimates

Term	Estimate	Std Error	t Ratio	Prob>|t|
Intercept	1.026	0.0058	175.8	0.0000
Time (Years)	0.0058	0.0015	4.0	<.0001

Plot 6a
Quantal response model 1%/year increase in R after 2 years using Time In-Service

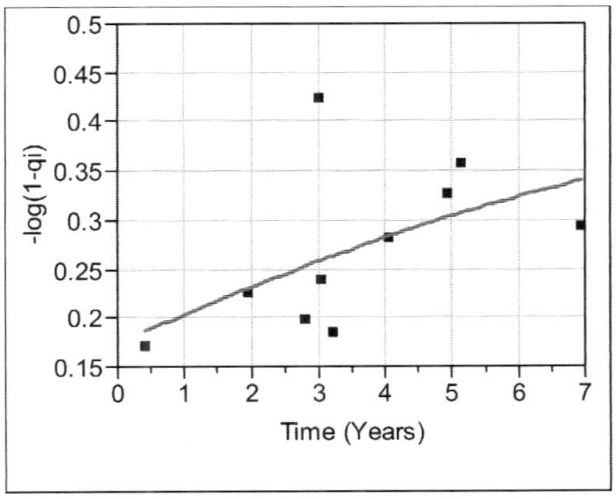

Plot 6b
Quantal response model 1%/year increase in R after 2 years using MCS Time

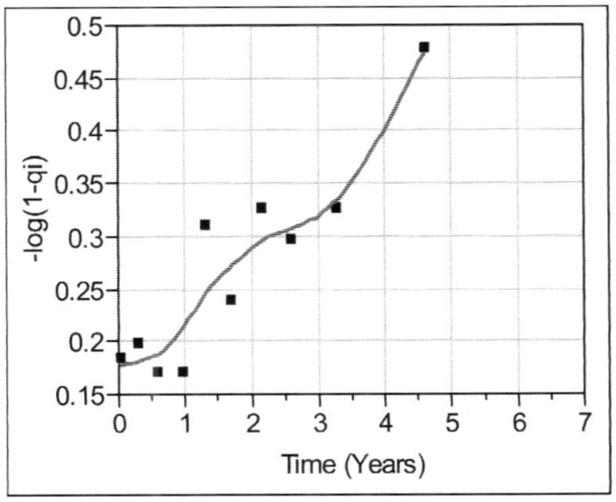

Plot 7a
Quantal response model 2%/year increase in R after 2 years using Time In-Service

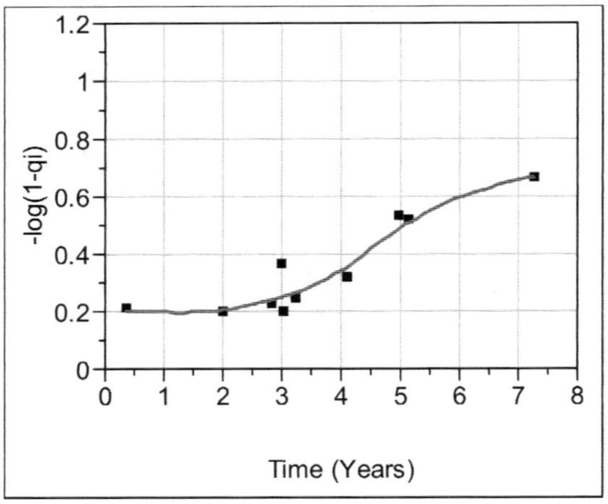

Plot 7b
Quantal response model 2%/year increase in R after 2 years Using MCS Time

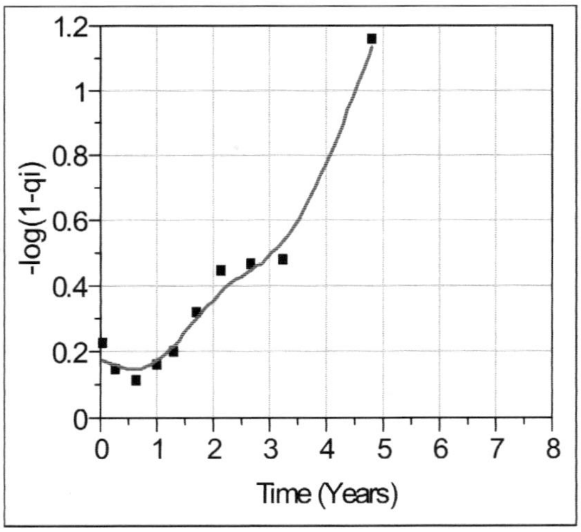

More Lessons "Re-Learned" from Corrosion Under Insulation

Tim Overton, The Dow Chemical Company

Abstract

Mechanical Integrity programs have been an essential element of process safety programs in the chemical and petroleum industries for decades. This is an area where considerable materials have been published regarding industry best practices for inspection and maintenance – including excellent advice on inspection for "corrosion under insulation". However, The Dow Chemical Company experienced a significant near miss event at one facility recently caused by lapses in understanding by local personnel of corrosion under insulation, as well as inadequate leveraging of learnings from plants of similar design.

This paper explains the circumstances behind this event and the key lessons Dow learned to help other companies avoid similar occurrences.

Before the incident

Dow's process safety performance historical performance has been good. No multiple fatality incidents since 1981. Both Injury/Illness statistics and normalized process safety incident performance are among the industry's best. Dow had corporate-wide mechanical integrity programs in place since the 1960s – and the mechanical integrity program standard called upon piping inspections consistant with API 570 since 2001 for piping in highly hazardous services (earlier standards existed that may not have been aligned to API 570).

The company experienced a few line failures resulting from corrosion under insulation (CUI); however, most of these failures were in piping systems not meeting Dow's previous hazard criteria within the mechanical integrity safety standard. However, some less severe incidents and a few close calls had occurred. Dow updated its corporate mechanical integrity standard in mid-2004 to clarify the mechanical integrity program's requirements and accountabilities. (The need for this will be described in more detail later in this document). Following the rollout of Dow's new mechanical integrity standard, there were indications that CUI inspection quality improved – including better piping systems inspections. A number of piping systems which were inspected after 2004 revealed corrosion under insulation which were fortunately corrected before failure.

The incident – the failure of a 8 inch carbon steel, high pressure hydrocarbon line due to Corrosion Under Insulation

On March 13, 2008, at a Dow's LHC plant, an 8 inch, schedule 20, carbon steel hydrocarbon line failed. The event began with a small pinhole leak, detected by the operators. While the operators were isolating and de-pressuring the line, it failed catastrophically. Fortunately, this catastrophic line failure produced such explosive force that the line bent in two locations – essentially sealing itself and avoiding a major release.

Fig. 1 - photo of the failed 8" pipe

– photo of pipe once removed

Fig. 2 – photo of the crease in pipe self-sealing the leak

The piping was a 30 year old, schedule 20, carbon steel regeneration line to a cracked gas dryer of an ethylene plant (one of two parallel drying trains). The material in the line was non-corrosive hydrocarbon cracked gas (40% ethylene). Due to the regeneration service, the piping alternates between three different conditions, and the location of failure was where operating temperatures cause frequent or continuous condensation and re-evaporation of atmospheric moisture:

Regeneration condition (4 bar, 220° C)
On stream time: 20 days (30 bar, -17° C)
Regeneration time: 24 hrs (220° C)
Standby: 19 days (ambient temperature)

Figure 4 – photo showing installed piping when out of service

Root Cause Investigation findings / Lesson's Learned

A number of root causes were identified in the root cause investigation from this incident. Some of the root causes were related to insufficient previous inspection. As noted above, Dow rolled out an update to their global mechanical integrity safety standard in mid-2004. This is Dow's current global mechanical integrity safety standard. Ironically, this line had last been inspected just a few weeks earlier in 2004. Therefore, the inspection organization had not yet been notified or trained on the new standard. Some of the root causes or lesson's learned described below were corrected in the new standard.

Cause #1 - Inadequate Inspection in 2004 and earlier

Upon reviewing the inspection records, and discovery of the amount of corrosion under insulation at the point of failure, it was clear that neither the inspection in 2004 nor any of the earlier inspections, had been performed in accordance with API 570. Specifically, insufficient insulation was removed to allow adequate external inspection. Instead, only a small plug of insulation at the elbow was removed, under the assumption that the elbow was the most critical inspection location.

Lesson Learned #1: Pre-2004 Mechanical Integrity Standard was too vague and left the burden to obtain external standards upon the local inspection and maintenance organization. The pre-2004 standard simply stated:

- ....piping systems in **highly** hazardous service shall be identified and registered *[i.e., the steps required to place in the mechanical integrity system for future inspections]* An evaluation shall be made to determine if a potential spill under normal operating conditions could exceed threshold external authority reportable quantities (e.g., EPA[13]). If the piping system has the potential to release highly hazardous chemicals in quantities greater than the threshold limit, then the piping system shall be registered.
- ... API 570 shall be used for guidance in performing inspections and repairs.

This wording left too much discretion to the plants to identify which piping should be inspected, and assumed that the plant or inspection organization would obtain copies and follow API 570. The pre-2004 standard also was vague in describing who was responsible for the steps of registration, inspection, and follow-up associated with the inspection program.

The 2004 version of the Dow mechanical integrity safety standard (rolled out a few weeks after the last time this piping was inspected), provided much greater clarity regarding piping to be inspected, insulation to be removed, and CUI susceptible regions per API-570. Examples of figures or text from the 2004 standard are below:

Figure 5 – Table describing different types of hazardous service and Dow's minimum piping inspection frequency

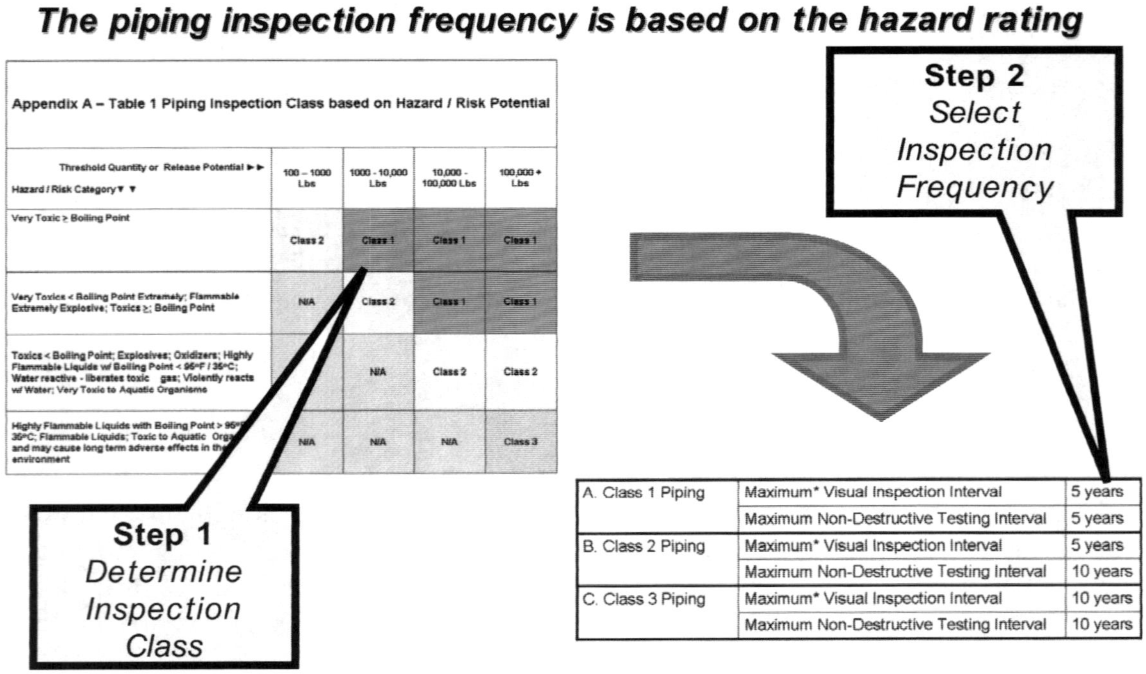

Appendix A – Table 1 Piping Inspection Class based on Hazard / Risk Potential

Threshold Quantity or Release Potential ►► / Hazard / Risk Category ▼ ▼	100 – 1000 Lbs	1000 - 10,000 Lbs	10,000 - 100,000 Lbs	100,000 + Lbs
Very Toxic ≥ Boiling Point	Class 2	Class 1	Class 1	Class 1
Very Toxics < Boiling Point Extremely; Flammable Extremely Explosive; Toxics ≥; Boiling Point	N/A	Class 2	Class 1	Class 1
Toxics < Boiling Point; Explosives; Oxidizers; Highly Flammable Liquids w/ Boiling Point < 95°F / 35°C; Water reactive - liberates toxic gas; Violently reacts w/ Water; Very Toxic to Aquatic Organisms	[illegible]	N/A	Class 2	Class 2
Highly Flammable Liquids with Boiling Point > 95° 35°C; Flammable Liquids; Toxic to Aquatic Org and may cause long term adverse effects in th environment	N/A	N/A	N/A	Class 3

A. Class 1 Piping	Maximum* Visual Inspection Interval	5 years
	Maximum Non-Destructive Testing Interval	5 years
B. Class 2 Piping	Maximum* Visual Inspection Interval	5 years
	Maximum Non-Destructive Testing Interval	10 years
C. Class 3 Piping	Maximum* Visual Inspection Interval	10 years
	Maximum Non-Destructive Testing Interval	10 years

Fig 6 – Extent of Corrosion Under Insulation Inspection Following Visual Inspection described in Dow's 2004 Mechanical Integrity Safety Standard

Piping Class	Approximate amount of CUI Inspection by NDE or Insulation Removal at "Suspect Areas" on piping systems within "Susceptible Temperature Ranges."	Approximate amount of follow-up examination with NDE or Insulation Removal in areas with Damaged Insulation on piping systems within "Susceptible Temperature Ranges."
1	50 % of suspect areas	**75 % of area of damaged insulation**
2	33 % of suspect areas	**50 % of area of damaged insulation**
3	**10 % of suspect areas**	**25 % of area of damaged insulation**

Susceptible Temperature Ranges – API 570 Piping Inspection Code, Section 5.3.3.1 e, f & h:

e. Carbon steel piping systems, including those insulated for personnel protection, operating between 25°F–250°F (–4°C–120°C). CUI is particularly aggressive where operating temperatures cause frequent or continuous condensation and re-evaporation of atmospheric moisture.

f. Carbon steel piping systems that normally operate in-service above 250°F (120°C) but are in intermittent service.

h. Austenitic stainless steel piping systems operating between 150°F–400°F (65°C–204°C). (These systems are susceptible to chloride stress corrosion cracking.)

Suspect Areas – API 570 Piping Inspection Code, Section 5.3.3.2

The areas of piping systems listed in 5.3.3.1 may have specific locations within them that are more susceptible to CUI, including the following:

a. All penetrations or breaches in the insulation jacketing systems, such as: 1) Deadlegs (vents, drains, and other similar items), 2) Pipe hangers and other supports, 3) Valves and fittings (irregular insulation surfaces), 4) Bolted-on pipe shoes and 5) Steam tracer tubing penetrations.

b. Termination of insulation at flanges and other piping components.

c. Damaged or missing insulation jacketing.

d. Insulation jacketing seams located on the top of horizontal piping or improperly lapped or sealed insulation jacketing.

e. Termination of insulation in a vertical pipe.

f. Caulking that has hardened, has separated, or is missing.

g. Bulges or staining of the insulation or jacketing system or missing bands. (Bulges may indicate corrosion product buildup.)

h. Low points in piping systems that have a known breach in the insulation system, including low points in long unsupported piping runs.

i. Carbon or low-alloy steel flanges, bolting, and other components under insulation in high-alloy piping systems.

j. Locations where insulation plugs have been removed to permit piping thickness measurements on insulated piping should receive particular attention. These plugs should be promptly replaced and sealed. Several types of removable plugs are commercially available that permit inspection and identification of inspection points for future reference.

The 2004 standard also describes all activities and responsibilities associated with the mechanical integrity program by role (e.g., Site Maintenance Leader, Equipment owner, Site Leader, Inspector, Designer) – to increase the understanding and accountability.

Including specific expectation and reproducing the requirements inspectors should follow from external standards are critical for consistant global mechanical integrity inspections.

Lesson Learned #2 – Mechanical Integrity Programs can not rely solely upon contract inspectors and need adequate involvement of plant personnel familiar with plant operations.

As is a growing trend within the industry, this facility had contract inspectors performing most mechanical integrity inspections. There was also inadequate communications with the plant operations personnel when developing the inspection plans. As a result, the inspector was unaware of the intermittent operations at different temperatures (hot, cold, ambient) which led to accelerated corrosion under insulation.

Effective communications and involvement of persons familiar with all modes of operation within the plant is critical to develop a robust inspection plan.

Cause #2 - Inadequate Leveraging of Learnings

During the root cause investigation, the team concluded that there had been indications that this specific pipe location was susceptible to CUI and the piping location should be: 1) replaced with thicker wall pipe or alternative metallurgy and 2) periodically inspected thoroughly for CUI (i.e., insulation removed). Examples include:

- This plant was a typical of a specific technology licensor's design, purchased from a previous operating company. A plant of identical design was owned and operated immediately adjacent to this plant by the other company. (The two companies shared a control room and maintained good communications.) The other company had identified significant corrosion at this specific piping location in their sister plant and had replace the piping.
- Dow also owned and operated other plants of a similar design in at least one other location. That plant had discovered CUI at this specific piping location and had replaced the piping.
- The inspection of the parallel train dryer in this specific plant had revealed excessive corrosion and had replaced the piping.

Yet none of these events prompted the plant to identify this specific piping location as a CUI susceptible region, accelerate the CUI inspection, remove more insulation for CUI inspections, or replace this specific piping location.

Lesson Learned #3 – An effective program to leverage learnings from other incidents should include discoveries of excessive corrosion or any other losses of mechanical integrity.

Following this incident, the technology experts in each of Dow's major business units were instructed to compile a "high priority CUI susceptible location" document for their technology based upon historical inspections and/or failures. These documents were distributed to plants around the globe with instructions to ensure that all such piping locations had completed a thorough CUI inspection. (Prompt inspection was recommended for any locations that had not been recently inspected.)

Other company actions following this event

Dow's leadership and employees were aware of the potential severity of this CUI incident. If not for the good fortune of the pipe bending in a perfect angle to self-seal, there would have been a very large discharge of highly flammable vapors and a high probability of a vapor cloud explosion resulting in major damages and potential human casualties. This awareness prompted Dow's leadership and company mechanical integrity program steering team to launch a number of corporate-wide initiatives. These included:

- A mandatory refresher training for all site and production plant leaders on the key elements and responsibilities under the corporate mechanical integrity program.
- The creation of a list of high priority CUI inspection locations, by technology, to be inspected as soon as possible.
- The creation and delivery of new CUI inspection training to train maintenance and operations personnel how to better identify areas susceptible to CUI.
- The launch of an additional, one-time mechanical integrity program audits of all major sites. (The mechanical integrity program of each site should be audited as part of the sites' ongoing periodic EH&S audits . This special audit was intended supplement this auditing with a highly visible site-wide, mechanical integrity focused audit to respond to this near miss event.)

Evidence that the increased emphasis is working

Since the rollout of the CUI special emphasis program a number of CUI discoveries have been reported. This is similar to what Dow saw in response to the rollout of the 2004 corporate mechanical integrity safety standard. This demonstrates there is increased awareness and emphasis in mechanical integrity is indeed avoiding future incidents.

Conclusion

Effective management of CUI risks is a critical part of proper process safety hazard management leading to good safety performance. There are so process safety management systems and best practices that it would be impossible to rank one as "the most important" to incident elimination. An effective mechanical integrity program would certainly be among the top. Justifying the expense of an effective CUI inspection program is often challenging since to identify all areas of CUI requires a very expensive process of removing insulation and re-insulating – even if the inspection reveals no corrosion damage. But the failures prevented by an effective CUI inspection program are often the most dangerous – since they can occur in areas where there was no previous indications of a problem.

Implementation of a Global Mechanical Integrity Program at Solutia

F. Russ Davis
Solutia, Inc.
575 Maryville Centre
St. Louis, Mo 63141

ABSTRACT:

This paper examines the challenges of defining a global mechanical integrity (MI) program for a chemical company with manufacturing facilities in North America, South America, Europe and Asia. The paper examines the differing regulatory requirements and engineering standards and also investigates the challenges associated with limited technical resources and limited availability of state of the art non-destructive examination methods in various regions. The paper further discusses cultural views of mechanical integrity in various global areas.

INTRODUCTION

Solutia is a 1997 spin off from Monsanto Chemicals. At the time of the formation of Solutia Inc. Monsanto was partner in a joint venture with Akzo Nobel called Flexsys. The Flexsys JV became part of Solutia. In 2007 Solutia bought out the Akzo portion of the JV and Flexsys became a fully Solutia owned organization.

Solutia is a widely diverse global company which creates many opportunities and challenges in the development and integration of a Global MI program. This paper will explore many of the challenges. While it is fully recognized that there is no one size fits all MI program, the basic concepts are the same no matter where in the world Solutia operates chemical plants and manufacturing facilities. The goal for all facilities is the same; to have a safe and reliable process. Equipment and processes must be designed to operate reliably. Good engineering practices must be recognized and followed. Process equipment must operate when needed; pumps have to move the process, vessels have to safely store, separate and mix hazardous chemicals. Safety critical instrumentation such as interlocks, alarms and controls must function on demand.

This paper will discuss some of the challenges and opportunities we have found in the implementation of a Global MI program. The challenges include:

- Determining what the company wants the MI program to be.
- Specific country regulations- compliance versus reliability. We do what we are required to do.

- Cultural differences. Not just different countries but also legacy MI programs.
- Lack of MI Reference standards globally
- Language differences

ESTABLISHING THE GLOBAL MI REQUIREMENTS

Solutia is a US based company and many of the MI program requirements are based on best practices developed in the US over the past many years. While some elements of a reliability program have been around for many years, the concept of a fully encompassing program affecting all process equipment and mitigating systems was promulgated in the US with the OSHA PSM standard 1910.119. In the European Union, Seveso II requires an establishment take all measures necessary to prevent a major accident and to limit their consequences. Organizations should determine precisely how they will achieve compliance with local laws and regulations. In many other countries such as Malaysia, China and Japan, laws addressing control of major accidents occurrence in the chemical and refining industries resemble the Seveso II and OSHA PSM standards.

The MI program at Solutia is viewed as a reliability program. The need to be in full compliance with all regulatory requirements in the host countries where Solutia has manufacturing establishments is recognized. An MI program focused only on compliance with the varied laws and regulations does not constitute an encompassing proactive MI reliability program. An MI program should be fully proactive. A proactive MI program consists of inspection, testing, preventative maintenance and predictive maintenance elements designed to identify equipment anomalies prior to failure or prior to demand. By identifying problems before failure, corrective measures can be planned and implemented to ensure the continued safe operation of critical equipment.

A Global organization should have a consistent enterprise MI program to direct the efforts and set the requirements for each of its entities. The enterprise program must fully define the expectations for all facilities. A global MI requirements program should designate all the equipment that will be covered. Covered equipment should include at least:

- Pressure vessels
- Piping
- Storage tanks
- Pressure relieving systems
- Emergency venting systems
- Safety critical controls
- Interlocks
- Pumps and other critical rotating equipment

The corporate MI requirements documents should designate the expectations particular to the individual equipment types. Facilities must designate who is responsible within the organization for MI. In most facilities the maintenance group is responsible for the implementation of the MI program. In many countries regulatory bodies conduct periodic inspections of specific equipment, generally pressure vessels, fired boilers and furnaces. The interface with the regulatory body may be handled through the Environmental, Safety and Health group.

The requirements of an MI program must start with the design and construction of the equipment. The corporate MI requirements program can specify the need for each facility to designate the codes and standards applicable to each equipment type. It should be the responsibility of the local establishments to reference the laws, codes and standards it must comply with for construction and installation of equipment into the sites' MI standard. This paper will discuss later the issues of compliance to each country's laws and regulations when designating engineering standards for manufacture of differing equipment types.

The enterprise requirements program should define basic inspections and testing methodology that each establishment must include in their facility MI program. The basic testing methodology may include visual internal and external inspection of pressure vessels and storage tanks. Thickness testing of pressure vessels, tanks and piping should be performed to establish corrosion rates and for evaluating the equipment suitability for continued service. The enterprise program may designate loop testing for interlocks and alarms. The enterprise MI requirements document should require that each plant establish written procedures for the testing and inspection of each equipment type. Specific testing protocols should be written as part of the local establishments MI procedures. Each facility must have a system for documenting inspections and tests and a system in place for the analysis of the data. Results gleaned from the data should be used for planning equipment repair and replacement.

The enterprise MI program should require that each of its facilities have a system in place for addressing inspection and testing results that are outside acceptable limits. In many cases establishments that have a MI program focused on compliance with regulatory requirements inspection results are not being used for planning future repair/replacement of equipment.

It is vitally important that an enterprise MI program is developed at the corporate level. The corporate program should clearly define the MI goals and expectations and direct each facility to implementation the program. The organizational expectations must be clearly defined at a level above the individual establishments, plants and facilities. A global MI program must direct compliance with all local laws but a MI program based strictly on compliance will fail to function as a proactive reliability program. The primary benefit of a comprehensive organizational MI program is reliable, safe process equipment and safety systems.

MI LAWS & COUNTRY REGULATORY REQUIREMENTS: COMPLIANCE VS PROACTIVE MI PROGRAM

One of the challenges of a global MI program is the many laws and codes applicable to each country. Establishments will strive to be in compliance with the local laws and regulations especially when inspections are being performed by regulatory inspection agencies.

In the US, steam boilers are licensed by the states and there are boiler regulations requiring inspections to be performed by approved inspection agencies and by state licensed and approved inspectors. These inspections must be performed at a frequency established by the states themselves. Reports are submitted to the state boards and licenses are issued for continued use of the boiler. Many other countries have similar laws for boilers and pressure vessels. OSHA through the Process Safety Management standard 1910.119(j) and the EPA through 40 CFR part 68 Risk Management standards require "covered" facilities to have an MI program. There is no requirement that the inspections be performed by state or federal regulatory agencies.

In the European Union, member countries are required to conform to 1999/92/EC, also known as ATEX 137, and to 94/9/EC also known as ATEX 95, and enact into laws these standards. EU standard 1999/92/EC requires under article 8 "that the workplace and work equipment, including warning devices, are designed, operated and maintained with due regard for safety". Standard 94/9/EC addresses "Equipment Intended for Use in Potentially Explosive Atmospheres". The 94/9/EC requirement is directed primarily at equipment "which are capable of causing an explosion through their own potential sources of ignition". The EU has also issued the PED/Pressure Equipment Directive (97/23/EC). Pressure equipment by the PED definition means vessels, piping, safety accessories and pressure accessories. The PED is not as technically prescriptive as the ASME VIII div I but it does require the use of accepted good engineering codes and practices. The EU members make use of the EN standards for engineering specifications as well as standards developed by organizations and regulatory bodies within the each individual EU country state.

In Italy, there are regulations addressing equipment under pressure including pressure vessels, pressure piping, and relief valves associated with pressure vessels. There too, local authorities perform inspections to allow continued operations of the pressure equipment. Inspection personnel are accredited by Italian authorities.

French regulations require establishments to have an operating permit which requires MI programs for pressure equipment and compliance to the ATEX (Atmospheres Explosibles) regulations for safety instrumentation and pumps operating in potentially explosive atmospheres. Inspection personnel must be accredited by the French accreditation agency COFRAC.

In the UK, several regulations address different aspects of MI. COMAH (Control of Major Accident Hazards), PSSR (Pressure Systems Safety Regulations), ATEX, and the PED (Pressure Equipment Directive) address construction, installation, and maintenance to ensure suitability for continued service. Inspection agencies and inspectors are certified by various UK certification bodies.

In Germany, the Regulation of operating reliability requires that pressure equipment, storage tanks, pressure piping, safety instrumentation and relief valves are inspected and tested to ensure suitability for continued service. The Betriebssicherheitsverordnung (BetrSichV) requires inspections to be performed on pressure containing equipment by competent persons at a frequency set in section 15 of the BetrSichV.

In Belgium, under the 96/82/EC, Control of Major-Accident Hazards Involving Dangerous Substances, the Dutch Federal Public Service Employment, Labour and Social Dialogue has issued laws that make it a mandatory legal requirement to have a list of critical equipment; this equipment should be controlled on a regular basis by a specific "control organism". Since Belgium is a founding member of the EU they do follow EU regulations such as the 1999/92/EC, 94/9/EC, and the PED/Pressure Equipment Directive (97/23/EC).

In China, the local regulatory agency reviews thickness readings and performs site inspections of pressure vessels to permit continued service. Inspection agencies and technicians must be licensed by the regulatory agency in order to perform the thickness inspections. It appears that pressure vessels are the only equipment for which regulatory agencies require compliance inspections.

In Malaysia, regulations promulgated under the Factories & Machinery Act & Regulations require MI programs to address pressure vessels, piping, and storage tanks as well as fire protection systems. Safety critical instrumentation, pumps and relief valves are not covered by regulations. Inspectors are certified by outside agencies such as ASNT (American Society of Non-destructive Testing, and CSWIP (Certification Scheme for Personnel).

The Japanese have laws regulating a mechanical integrity program for chemical and manufacturing facilities. The Japanese Laws including the High Pressure Gas Safety Law, the Fire Fighting Law, and the Industrial Safety and Health Law require adherence to Japanese Industrial Standards. The Japanese International

Center for Occupational Safety & Health (JICOSH) Ordinance on Safety of Boilers and Pressure Vessels requires manufacturing, installation and management inspections. These inspections are required to be performed by certified inspectors by the Ministry of Health, Labour and Welfare. JICOSH has also issued guidelines on the construction of electrical equipment for explosive atmospheres. By Japanese law, owners of chemical plants are required to manage risks associated with the equipment used in the plant.

Mexico has issued MI requirements through the Secretariat of Labor and Social Security. The Health and Safety Norms (NOM) issued by the Secretariat contain some inspection requirement especially related to pressure vessels and boilers such as NOM-020-STPS-2002. Government approved inspectors perform pressure vessel inspections, approve or disapprove the vessel for continued use. Ordinances such as NOM-026-STPS-1998 Heath and Safety Colors and Signs and Identification of Risks of Fluids in Pipes and NOM-005-STPS-1998 Health and Safety Conditions in the Workplace with Regard to the Handling, Transport and Storage of Dangerous Chemicals require facilities to conduct periodic visual inspections of the installations and structural elements and to document these inspections. When breakage, cracking, warping, material fatigue, deformation, collapse or other similar conditions are detected a report must be filed and repairs carried out to correct the damage.

In Brazil, NR-13 is the specific legal requirement for boilers and pressure vessel MI programs. NR-13 requires inspections of boilers at least annually. However, NR-13 does not address storage tanks, rotating equipment, piping or other equipment considered safety critical equipment.

Laws, standards and local regulations associated with a MI program are varied and disconnected for many countries. The prevalent theme for most countries is pressure vessel and boiler mechanical integrity. While pressure vessels and boilers must operate safely they do not equate to all of the equipment that must be included in a complete proactive MI program. Piping failures statistically are more frequent than vessels yet many countries do not require piping to be included in the regulated MI program.

The following countries have mechanical integrity regulatory requirements to some level or another:

- USA
- Belgium
- Germany
- UK
- Mexico
- Brazil
- Japan

- Malaysia
- France
- China
- Italy

The EU countries have enacted regulatory guidance for the manufacture and sale of equipment for use in chemical, refining, and petrochemical establishments. Many other countries have not addressed all equipment that should be included in an MI program. The MI program must ensure all critical equipment functions as designed on demand yet there is definitely a gap between regulatory requirements and an encompassing MI program for many countries. The EU countries and the US have comprehensive process safety regulations addressing the requirements for maintenance inspections of "covered" equipment. Mexico and Brazil do not appear to address maintenance inspection requirements other than pressure vessels and boilers in a comprehensive regulatory manner.

MI REFERENCE STANDARDS AND RECOMMENDED PRACTICE

Mechanical integrity consensus guidelines are a valuable tool for every MI program. Those responsible for ensuring the integrity of process equipment need accepted good engineering practice on the design, installation and maintenance of the equipment. Consensus standards for detailed design and inspection can and do provide guidelines for implementing MI programs. In the US, the American Petroleum Industry (API) write and publish construction standards, recommended practice guidelines and inspection guidelines for just about every equipment category that will be included in an MI program. While many other countries have regulated or accepted design standards, few of these countries have maintenance standards like the API inspection and repair standards for pressure vessels, storage tanks, and piping. In the US, ANSI/ISA S84.00.01, Functional Safety: Safety Instrumented Systems for the Process Industry Sector, and its EU equivalent IEC 61511 give guidelines for safety instrumented system (SIS) design and maintenance.

When a company determines a global MI program is in its best interest for achieving safe, reliable equipment for their operating establishments, global MI requirements standards must be developed to provide guidance on the design, inspection and maintenance of equipment covered by the MI program. The consensus standards currently in use in the US are making progress in becoming internationally acknowledged good engineering practices. The API is currently transforming many of its standards into ISO specifications through the ISO/TC67 standards committees. Thus far, the API/ISO work has focused primarily on the oil and gas production standards. However, a few of the standards applicable to chemical plant MI programs have been issued as ISO standards.

The lack of global standards that address the different equipment types included in an MI program can create a problem. A global MI program should be consistent across the company with the understanding that local establishments must be in compliance with the laws and regulations of the host country. In the corporate global MI standard, requiring compliance to a specific design code may have to be avoided. It may be more appropriate for the corporate standard to require that equipment first comply with the laws and required design codes of the host country and then to the reference codes and specifications that the corporate company has accepted as good engineering practice.

A company should determine what codes and standards the organization accepts. The global MI requirements procedures should then require each of the manufacturing plants to adhere to the accepted standards when there is no conflict with local laws. Each plant must use the standards and recommended practices chosen by the company as accepted engineering practices for design criteria and as references for maintenance work practices.

LANGUAGE CHALLENGE

At first glance language appeared to be a challenge that could derail a global organization MI program. This does not appear to be quite the challenge that it was initially thought. English as a second language is fairly prevalent in many parts of the world. In some cases, recognized good engineering standards have been translated into other languages. It is important for the MI program manager at the different facilities be able to speak English. Benchmarking, networking and shared services must be done in a common language. In the world today the most common business and technical shared language is English. For an American organization corporate management should consider the language issue as part of its hiring practice for each sites MI responsible position. English language training should be incorporated into the individual business plan of those involved in managing the MI program at each different location.

CULTURAL AND LEGACY CHALLENGES

When different manufacturing establishments in different countries are queried about the MI program for their local facility the response is "We are in compliance with the laws, ordnances and regulatory requirements". In those countries whose laws and regulations require inspections by the government or by third party inspection companies authorized by the government the focus of the MI program will typically be to ensure compliance. The countries that have laws and regulations requiring specific inspections do not require inspection of all the equipment that should be included in the MI program. The local establishment's culture tends to accept the inspections performed by the governmental authorized agencies as the complete inspection requirements for the equipment. Typically, the regulatory inspections validate that the equipment is suitable for service at the time of the inspection. These inspections do not calculate a corrosion rate or determine anticipated vessel life based on the corrosion rate.

In the US, states require inspections by authorized agencies of fired boilers and steam generators. These inspections are directed at the safe operations of the steam boiler at the time of the inspection. Fired boilers and steam generator regulatory inspections do not anticipate damage based on corrosion rates so that repairs and replacements can be anticipated and planned for in advance of failure. This type of focused "safe to operate" inspection seems to be the regulatory norm for countries such as Japan, China, and Malaysia.

Legacy MI programs have been another challenge noted in the implementation of a global MI program. Establishments tend to comply with the local country laws and then to the requirements of the owing corporate entity. When acquisitions occur the new owner has inherited the past owners MI program. Every organization has their concept of a MI program, some better than others. A gap analysis should be performed to recognize any opportunities for improvement. Roll out of the corporate MI program requirements and training of personnel involved in implementing the program should be done within a realistic time frame. In some cases there may be the need to retrain personnel whose MI programs have been fixed strictly on compliance.

PATH FORWARD

A proactive mechanical integrity program must be defined at the top levels in the company. Corporate leadership must clearly state expectation for each establishment and facility within the corporation. A written MI document should direct each plant on how to achieve the corporate expectations. At the same time, the corporate MI program must take into consideration local laws, ordinances and regulations. Reference standards and codes should be noted by the corporate MI program. Corporate engineering standards should set the standard for all establishments, again considering corporate engineering standards should not be in conflict with host country requirements.

Implementation of an encompassing corporate MI program has to be followed with a plan for implementation at each local establishment. The plan should be developed by each site. MI audits should be performed to ensure compliance to the plan and to the requirements of the corporate MI program.

Metrics must be established to evaluate compliance to the intent of the corporate MI program. The percentage of inspections performed on time, as scheduled, for equipment "covered" by the MI program is an accepted leading MI metric. A metric should also be established to measure the effectiveness of the MI program. Plant lost production time caused by equipment failures or the number of emergency work orders as a percent of planned work orders is a valid measurement of the effectiveness of an MI program.

Management should readily see the benefits of an effective global MI program through reduction in lost production time due to equipment failures. Maintenance costs are reduced when a proactive MI program is fully integrated into the plant business model, and planned work is far cheaper to perform than emergency work.

The primary benefit of a global mechanical integrity program is safer plants. Failures in MI programs have led to catastrophic failures. Plant equipment must operate safely and when needed. Vessels and piping systems must maintain containment of hazardous chemicals and energy and safety equipment such as relief devices and critical instrument must function when needed.

47e Adoption of Mechanical Integrity Principles at Air Liquide Large Industries US

Danny Ledbetter
Air Liquide
Xxx, TX 00000
Danny.Ledbetter@Airliquide.com

Abstract:

This information is offered for presentation at the Mechanical Integrity session of the Plant Process Safety Symposium. The purpose of this abstract is to describe the execution plan Air Liquide Large Industries US (ALLIUS) employed to implement Mechanical Integrity Principles in North America and Globally. The majority of ALLIUS processes are considered non-covered, as defined by OSHA 1910.119, so the decision to advance Mechanical Integrity Principles was not a requirement, instead it was to improve existing and future assets. The effort started by getting executive management engaged and onboard with the goals and objectives of the program. Upon approval, a Mechanical Integrity leader was selected and empowered to cultivate the program. Business leaders supported the effort by identifying and prioritizing plants where the need was the greatest. Plant by plant assessments followed. Certified inspectors and experienced personnel knowledgeable in "Recognized and Generally Accepted Good Engineering Practices" (RAGAGEP) were used to conduct the assessments and identify asset deficiencies. Corrective action plans were developed on a plant by plant basis, funded, and tracked through resolution. The ALLIUS Mechanical Integrity effort did not stop at the plant level, but instead caused systemic changes in the existing preventive maintenance program as well as new global organizational roles. The new roles are to focus improvement in capital projects. Personnel are now dedicated to front end and value added engineering efforts. Adoption of Mechanical Integrity principles has aided ALLIUS to improve plant reliability through reduction of unplanned equipment failures.

Implementing Risk Based Inspection on a Global Scale

Andrew Morey
Rohm and Haas Company
Croydon, PA

ABSTRACT

In 2007 Rohm and Haas Company made the decision to implement risk-based inspection in over 70 covered facilities worldwide as the most cost-effective method of complying with our internal process safety management standard for mechanical integrity of pressure vessels, storage tanks, piping and relief devices. The company has not traditionally dictated how to implement company standards or programs. However, standardizing on a common software system and procedures is bringing benefits in work process improvements, reporting and auditing. The company realized the need to provide plants with technical support to implement and manage the new system and developed a shared-services model for providing technical support that minimizes resource requirements at the plant level. A unique work process was developed to enable implementation on aggressive schedule while minimizing the use of outside contractors. We are already seeing benefits from the implementation but realize that we face additional challenges in bringing together a complicated network of mostly small plants in diverse businesses, especially in difficult economic times.

INTRODUCTION

Safe operation of process plants is both a regulatory requirement and a sound business decision. The Rohm and Haas Company operates facilities throughout the world and has an internal process safety management standard that is intended to comply with regulatory requirements and ensure safe operation of these facilities. The company has about 70 facilities that are covered by the process safety management standard. Most of these facilities are small (50-150 employees) and 34 of the sites have a similar or identical design.

A significant component of a process safety management system is regular inspection and testing to determine the mechanical integrity of process equipment. Risk-based inspection (RBI) is a recognized and accepted methodology for prioritizing these inspections. The engineering analysis that leads to this prioritization can allow more effective inspections to be selected as well as focusing resources on inspections of equipment with the highest risk.

BACKGROUND

The British Petroleum, Texas City process safety incident in 2005 prompted many companies to take a closer look at process safety management systems. At Rohm and Haas, internal audits were conducted to compare how different facilities were implementing the corporate process safety management standard. The audits showed that there was a lot of variability between sites, especially in countries outside the United States. Since mechanical integrity is a key component of process safety, several facilities were benchmarked to compare their approaches. Mechanical integrity systems are traditionally built around time-based preventive maintenance (PM) inspections utilizing the same techniques for almost all equipment. Although most facilities used a common computerized maintenance management system (CMMS), it was difficult to assess the situation at most facilities remotely. Several differences were noted:

- Work-level procedures were different in most facilities
- Inspection intervals were not consistent
- Inspection documentation was highly variable
- No standard tools existed to facilitate reporting or auditing

The benchmarking team found that several facilities were having good success with a risk-based inspection system (RBI). The RBI system utilized an integrated software solution – the software manages the risk analysis tool as well as the inspection data and results. Many companies are now using RBI tools as a means of prioritizing and focusing inspection resources. However, the integrated software offered several additional advantages for the company:

- Standardized inspection planning process and tasks
- Standardized inspection intervals
- Standardized collection of inspection results and history
- Reporting tools for overall compliance and detailed auditing

TECHNICAL ISSUES

The advantages of RBI are becoming well known in the industry. An RBI study can result in more efficient, better targeted inspections. This can save money while maintaining a more effective program. However, implementing and maintaining an RBI program requires an up-front engineering study that requires technical resources. The program must be implemented and maintained by personnel knowledgeable in the inspection industry as well as the software program. In addition, process engineering and corrosion engineering knowledge is needed to ensure that the proper operating conditions and damage mechanisms are identified. Operating and maintenance experience is needed to help validate modeling and criticality results. It was recognized that these resources would be difficult or impossible to obtain in many of our smaller facilities.

CULTURAL ISSUES

As noted earlier, Rohm and Haas is a network of mostly small facilities lead by business-level teams. The company has not traditionally dictated how to comply with corporate standards or initiatives. Plants were free to choose how to meet the requirements of a standard. Because of this, most plants worked independently or in business level groups. This type of company culture can be useful to support innovation and fast reaction to business needs, but does limit sharing of best practices and standardization across the company. While the advantages of RBI and standardization were very attractive to the benchmarking team, it was also apparent that a different management model would need to be used to move the company forward in a new direction.

IMPLEMENTATION RESOURCE OPTIONS

Implementing a technical solution like RBI is commonly done by developing an in-house team of experts, hiring a contract firm or a combination of both. In fact, Rohm and Haas had used all of these methods in early RBI implementations. As the team considered a corporate-wide implementation, there were several issues that needed to be addressed:

- Technical resource limitations at smaller facilities
- Contractor costs for a "turn-key" job were considered too high.
- Travel costs for contractors or company resources were a concern given the many facility locations worldwide.

To address these concerns, the team developed an implementation process that uses shared technical resources along with local plant personnel and minimizes the use of contractors and travel costs.

SHARED SERVICES – SHARED RESPONSIBILITES

In 2007 Rohm and Haas made the decision to implement RBI software in all covered facilities. Company leaders agreed that the advantages of the RBI solution, coupled with a new implementation strategy were compelling enough to support moving forward with a common solution across all sites.

The new implementation model provides corporate technical support along with plant responsibility for managing the execution of inspections. In addition, plant personnel have a key role in data collection for the project implementation. This model balances the need for resource sharing with the need to maintain an appropriate level of ownership at the local level. It also minimizes the use of outside contractors as a means of controlling costs. The new roles are described in Figure 1.

Figure 1: Mechanical Integrity Roles

Corporate Coordinator

- Provides oversight and DB audits, training, networking

Inspection Risk Analyst

- Manages the program for a plant or multiple plants
- Can be either a Plant, Business or Corporate position
- Trained and knowledgeable in RBI software, inspection and basic corrosion

Mechanical Integrity Coordinator

- Plant position
- Owns the inspection schedule for the plant
- Interfaces with Production to schedule and execute inspections and document results
- Basic RBI software training (run reports and input inspection data, but no risk training)

Inspector

- Executes inspections
- Certified by API or other recognized organization
- Typically contracted

The corporate coordinator manages the overall implementation process as well as the ongoing program. This person is the owner of the software and the corporate mechanical integrity procedures. The coordinator is responsible for database auditing and overall reporting. Another key function is facilitating inspection risk analyst training, development and networking so that we develop and maintain this new competency. Qualifications include knowledge and experience in the inspection codes, mechanical integrity, basic corrosion and a thorough knowledge of the RBI software.

The inspection risk analyst is a key position in this model. This person manages initial implementation and then supports the ongoing maintenance of the program. In most cases, the inspection risk analyst supports multiple small plants. Larger facilities may have a dedicated inspection risk analyst. This person has experience in mechanical integrity and the inspection industry and is trained in the RBI software. The inspection risk analyst runs the criticality calculations, develops inspection plans and reviews inspection results. By staffing these positions, smaller plants do not have to develop and maintain these competencies for a part time role.

The mechanical integrity coordinator is the plant-level owner and contact for mechanical integrity. This person interfaces with production personnel and ensures that equipment inspections are executed and properly documented in the software. In most plants this is a part-time role. Qualifications include a basic knowledge of process plant equipment and general training in the RBI software. The coordinator runs reports and inputs inspection results but is not responsible for running the risk calculations or evaluating inspection results.

These positions provide the needed technical support without adversely impacting plant organizations. Because most of the company's facilities are small, using shared resources makes sense – mechanical integrity is only a part-time role for personnel in most plants. The functional relationships of these positions are illustrated in Figure 2.

Figure 2: Mechanical Integrity Functional Relationships

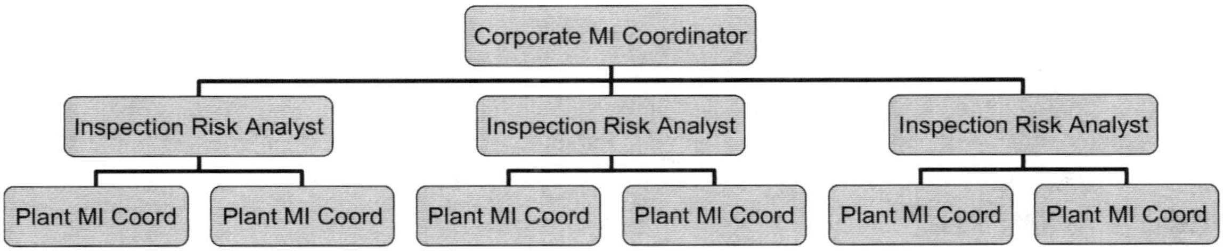

Two additional support roles are required for program implementation - corrosion engineering and process engineering. Corrosion engineering support is provided by the company's existing corrosion engineering organization. These resources are critical for ensuring that the proper damage mechanisms are identified during the program implementation and are also consulted for non-routine inspection plans and failure analysis. Process engineering support is provided by plant-level personnel and is necessary to verify proper operating conditions and review of the criticality results.

IMPLEMENTATION WORK PROCESS

The general project implementation steps followed are not unique and would be common to most RBI implementations. The major process steps are shown in Figure 3.

Figure 3: RBI Implementation Steps

- Create a covered equipment list
- Gather the required equipment and process information
- Conduct a Corrosion Study
- Determine initial Criticality and develop Inspection Work Plans
- Complete inspections (visual inspection, sketches and NDE) and evaluate results
- Update Criticality and develop next Inspection Work Plans

However, the way in which these steps are carried out has proven to be effective and efficient for a global implementation of small facilities. As discussed previously, the inspection risk analyst manages the implementation process and schedule. The majority of the project management is conducted remotely from the plant location. Most meetings with plant personnel are held via teleconference and web-meeting. Plant resources play a key role in executing the project efficiently, but are not required to take a leadership role. The following is a brief discussion of each project step:

Covered Equipment List

Each facility is required to develop a covered equipment list. This requires process knowledge to determine if a process falls under the company's process safety management standard. The inspection risk analyst will assist the plant with the development of this list as needed. Equipment that is not covered under the company standard but is covered by another regulatory requirement is generally included so that the MI system will be complete.

Data Collection

This is a key step as the quality of the RBI analysis depends on obtaining good information. Plant personnel are required to complete the primary data collection. Although this is a significant resource requirement, the schedule is set to allow the plant sufficient time to collect the data. In most cases the required data can be collected in about one man-hour per equipment. Data sheets were developed to streamline the process and to minimize the technical expertise needed by the data collector. Plant personnel with a variety of backgrounds can be quickly trained to collect the data.

Maintenance planners, production supervisors, operators, engineering technicians and administrative assistants have been used in addition to the more "typical" resources for a project like this such as a maintenance or reliability engineer. In almost all cases, a process engineer or experienced production supervisor will either supply or review the necessary process information to ensure accuracy. From experience we have found that the needed information is rarely located in one location. Most information is located in the equipment file (paper or electronic) but it is usually necessary to go to the equipment in the plant to verify information. For this reason, the datasheets are generally filled out on a printed form. This may seem counterintuitive in this age of computers, but we have found that the data collection actually takes less time using a paper system. Another advantage to this method is that the data collector does not need computer or typing skills. The data is obviously needed in an electronic form to load into the RBI software – this task is cost-effectively handled by using a contract data transcription service. This company takes the paper datasheets and transcribes the information into the required format. The transcription company's system includes format masking and error checking of key fields to improve data accuracy.

Corrosion Study

The corrosion study is a critical step in obtaining a quality RBI analysis. The inspection risk analyst prepares and leads the study after the data collection is complete. A corrosion study diagram is prepared by summarizing equipment and process data and adding the information to a process flow diagram. This information is supplied to the corrosion engineer prior to the meeting. The corrosion engineer provides vital information about expected damage mechanisms and corrosion rates, but plant experience is also very important for validation. The following positions should be represented at the meeting to ensure a quality study: inspection risk analyst, corrosion engineer, process engineer and one or more plant personnel with maintenance experience. The corrosion study is normally held as a web-meeting with a teleconference. Based on our experience, a series of two hour meetings are more effective than longer meetings. Small facilities with one process can usually be completed in one meeting; larger more complex facilities may require one or two meetings per process.

Initial Criticality and Inspection Work Plans

After the corrosion study is complete the data is combined for import into the RBI database. After import, an initial criticality is calculated. This represents the criticality or risk profile of the equipment based on the initial data gathering. Inspection plans are generated based on this criticality. The inspection risk analyst completes these tasks. At this point in the process, a site visit is planned to review the data results and train the plant MI coordinator. After the site visit, the set-up phase of the project is considered complete.

Complete Inspections

The plant MI coordinator is responsible for completing the required inspections. The amount of initial work depends on the availability and quality of existing inspection data. The primary focus is on completing on-stream visual inspections, thickness readings and corrosion-under-insulation inspections. The MI coordinator inputs the inspection results in the RBI database and notifies the inspection risk analyst for review of the results. Non-conformances are entered into the database for tracking as required.

Update Criticality and Develop Next Inspection Work Plans

After reviewing the inspection results the inspection risk analyst recalculates the criticality and plans the next inspection. Since the inspection strategies are based on the criticality, the next inspection interval and/or tasks could be different if the criticality changed. At this point the RBI implementation is considered complete but the cycle of inspections, evaluation, calculation of criticality and new inspection plans continues.

BARRIERS

Company leaders set an aggressive timeline for implementation. Our goal is to have all covered facilities using the RBI system by the end of 2010. The biggest challenge to meeting this schedule has been the availability of resources at the plant level. Small plant staffs are lean and we have had to adjust schedules to balance priorities with other plant or corporate initiatives. Another concern is the impact on plant budgets. Because of the shared services model, plants do not have to budget for inspection risk analyst support but they are responsible for all inspection costs. This is becoming more of a concern in light of our current economic situation. However, even with the budget and resource concerns excellent progress is being made. This would not be possible without strong support from company leaders.

CONCLUSION

Risk based inspection is an excellent tool for targeting and prioritizing inspection efforts. The benefits are becoming well known in industry, but the implementation process is often a barrier to more widespread use of the tool. By using a unique approach to implementation, coupled with strong leadership support, Rohm and Haas Company is taking advantage of the benefits of RBI to improve process safety at facilities throughout the world.

The Process Safety Guerilla: Giving Your Company a Future, Even if it Doesn't Want One

Scott Berger, Center for Chemical Process Safety
American Institute for Chemical Engineers
New York, NY 10016
646.495.1370
scotb@aiche.org

ABSTRACT

You understand well that a strong process safety program not only prevents incidents but also improves overall operational discipline, quality, efficiency, and profitability. Unfortunately, your management thinks that process safety is some kind of hippie environmentalist thing that will shave a few dollars off quarter-over-quarter growth. If only you could implement the kind of transformative program demonstrated by the Center for Chemical Process Safety (CCPS) founding companies and other industry leaders; unfortunately, you're not at a sufficient level to champion such a transition, and the people at that level aren't even in the ballpark, never mind willing to step up to the plate. Some day, maybe soon, the plant will go up in flames, and along with it will go your 401(k) and the livelihoods of your family, friends, and neighbors.

Now is the time for the downtrodden to rise up. How? By implementing the time-honed guerilla strategies used throughout the generations by those who lack the power of position. These strategies include theft, gangs, subterfuge, sabotage, arson, assassination, official corruption, and others.

Well... not really. But these "crimes" do suggest legal and effective ways of building a stronger process safety program from the bottom up. This paper will summarize ways that clever and persistent "process safety guerillas" have precipitated positive change at their companies.

Disclaimer:

This paper does not represent the views of the American Institute of Chemical Engineers, its Center for Chemical Process Safety, or for that matter the author. It is intended only as an attempt to look at the implementation of process safety management programs from a different perspective than usual. While terms are used that are suggestive of criminal activity, their use is only in the metaphorical sense, and this paper does not advocate illegal activity of any kind.

INTRODUCTION

The Center for Chemical Process Safety made a conscious decision in 1989 when it published its groundbreaking work "Guidelines for Technical Management of Process Safety" (1) to include as a critical technical element "Obtain top management support." By doing so explicitly, rather than by including process safety as part of a market-basket of engineering programs to be considered as a whole by top management, CCPS took a strong stand that process safety depends on culture and leadership as well as engineering excellence, and that it needs to be explicitly recognized as a key facet of operations.

Indeed, many companies have adopted CCPS' recommendations and work hard, even during the current economic downturn, to maintain top management awareness and support for their process safety programs. For example, a senior executive at one leading company, observing what some might have been considered an insignificant increase in process safety incidents at his company coupled with a negative trend in leading indicators, called an emergency meeting of plant and engineering managers to drive a proactive response before the increase could become significant. At a second company, another senior executive took similar action after he noted a trend in "lucky near misses"; incidents with minor consequences which could have had major consequences had they happened at a different time, or in a slightly different way.

But what if you don't work for such a company – rather, for one that:

- doesn't know about the process safety risks in its operations, or knows but doesn't rigorously manage them?
- is aware of its potential process safety risks and has a compliance-based program for which a few Health Safety and Environmental (HSE) professionals have responsibility, but doesn't have widespread support and participation by other employees?
- has a good occupational safety program that is not aware of process safety issues or regulations?
- handles both regulated and non-regulated substances which are not covered by process safety regulations and is in nominal compliance with respect to regulated materials but ignores non-regulated substances?
- top management focus has slipped?
- is completely clueless about process safety?

This paper explores an alternative approach to the recommended top-down strategy to implementing process safety, and is equally useful in establishing a top-down process for the first time and restoring a top-down process that has been lost. The anonymous anecdotes shared in this paper, by people who are quite typical of chemical engineers who have worked in process safety, were communicated personally and privately to me over the past fifteen years. Taken together, these stories bring to mind several historic figures who you would never think to associate with process safety.

THE STRATEGY OF THE PROCESS SAFETY GUERRILLA

Although the people who shared their anecdotes probably don't realize it, the approach they collectively describe follows Mao Zedong's basic principles of guerilla warfare, and to a surprising extent the methods explained in the Al Qaida manual. Mao (2) observed, "[Guerrilla warfare is] the inevitable result of the clash between oppressor and oppressed when the latter reach the limits of their endurance." While it may be fun to talk of industry senior management as oppressors, by all conventional measures this is hardly the case. In this case, we generally interpret oppression as simply the absence of strong support for process safety which arises not from malice but from ignorance or competing priorities.

Mao further observed that the oppressed, "lacking the quality of independence, are of themselves incapable of providing a solution to the struggle," unless they take up guerrilla action to help. Mao cites Lenin (3), noting that when people are oppressed, a "people's insurrection and a people's revolution are not only natural but inevitable." Given the natural motivation of chemical engineers to improve the process, the product, and the world, process safety guerrilla activity is therefore not at all surprising.

Mao's strategy for a process safety guerrilla action is based on the following key steps:

1. Arousing and organizing the people
2. Achieving internal unification
3. Establishing bases
4. Equipping forces
5. Gaining strength
6. Gaining control

A process safety guerrilla action requires leadership. In this scenario, this leadership has to come from someone who does not have leadership authority in the company. The leader or leaders of the process safety guerrilla uprising must be well educated in process safety, resolute in their dedication to preventing incidents, loyal to the cause and to their colleagues, sincere, and strong. Mao recognized that the leaders may not start with strong knowledge or leadership skills, but by trying, they will develop the knowledge and skills and prevail. Let's examine the application of Mao's basic strategy to process safety more closely.

IMPLEMENTING A PROCESS SAFETY GUERRILLA STRATEGY

Arousing and Organizing the People

The process safety guerrilla leader must draw supporters from those who are all around him. At a plant this could include other engineers, maintenance staff, and operators. How do process safety guerrillas do this? The most commonly reported way is via informal gatherings around the CCPS Process Safety Beacon posted by the coffee pot, tea pot, or water cooler. Another common way is to circulate articles about accidents at similar plants.

It is best to draw supporters from a range of roles, since a good process safety program requires such a broad diversity of skills. Over time, the leader must draw into the process safety guerrilla organization people with all the expertise needed to advance process safety. As the great guerrilla leader George Washington did by engaging the French, leaders can seek experts from outside the operational unit and even from those outside the company. Many respondents cited how they use their membership in CCPS and attendance at the Global Congress on Process Safety and its forerunner conferences to find advisors for their cause.

Achieving Internal Unification

A command and control structure is not necessarily required for a process safety guerrilla action. While the organization that Mao envisioned and ultimately amassed to drive the Japanese out of China was quite large and structured in intricate detail, other guerrilla groups, most recently Al Qaida, avoid command and control and instead create alignment around a core set of principles and standard responses to challenges that are presented. Either way, the first job of the leader is to overcome the natural tendency of people to say, "I am only a process engineer," "I am only an operator," or "I am only a mechanic," and get them to think, "I am a process safety guerrilla." This uniform direction of thought is necessary to assemble the loose group of people interested in *doing something* related to process safety, not into a formal team, but into an informal "gang."

To further strengthen the uniformity of thought, the gang should have a slogan, "colors" or a secret handshake to help identify themselves to each other. For example, a group of engineers whose companies had good programs, which were not, they felt, good enough, assembled a few years ago under the name "Process Safety Zealots." Another group agreed to wear corduroy slacks instead of the usual jeans or khakis to identify themselves. One advantage of having a standard form of dress is that it is easy to identify the other members of the process safety guerrilla gang. In the case of the corduroy wearers, the rustling noise created by the fabric rubbing as they walked also tended to remind workers that they were about to get safety feedback, so eventually the process safety guerrilla only had to walk nearby and the worker would correct his unsafe acts.

If E.I. DuPont was the first executive champion of process safety, then Trevor Kletz may have been the first guerrilla. As an assistant works manager, Kletz added a voluntary discussion of lessons learned from process safety incidents to his monthly staff meeting. His staff members were welcome to attend or not as they chose, and other works employees were also invited to join in if they chose. In writing of these meetings (4), Kletz focuses on the learning value of these sessions, diverting attention from their true nature as process safety guerrilla unification sessions. However, as the gatherings progressed, the topics evolved from overlooking blinds to bypassing interlocks to management of change to process safety culture, thereby revealing the true guerrilla nature of these meetings.

Establishing Bases

Once a core of supporters has been developed, the activities of this core can be expanded on by establishing other groupings that operate based on the same unifying principles established for the original group. In this way, the support for process safety can be spread to others within the plant, and ultimately within the company. The leaders for the new "gang" may come from within the original gang, or may be identified from the new targeted population. Engineers with corporate responsibility were reported to be good vehicles for spreading interest from one plant to another. The usual method involves the creation of an "interest group" or network of engineers across the company interested in process safety. It has become much easier to form such interest groups in the past few years due to the expansion of teleconference capabilities. Interest groups may start to look a lot like the officially sanctioned "Process Safety Leadership Teams" which exist in companies with top-down support, and getting to this point is, in fact, the goal of the next two steps.

Equipping Forces

A process safety guerrilla organization may not need much special equipment; however, it will need the various tools of the process safety trade, such as books, software, checklists, action item follow-up lists, etc. The guerrillas should share tools where possible (and since this paper advocates legal means, sharing should be subject to copyright and intellectual property law) so that common approaches can be used. One legal means shared by a respondent is that the purchase of corporate licenses to needed software, databases, and book libraries are distributed amongst individual plants represented in the network. Since for many products a corporate license doesn't cost significantly more than a site license, the process safety guerrilla organization can equip the company with the tools it needs without all the costs appearing in one cost center.

Some process safety guerrilla organizations have assembled their own tools, creating spreadsheets, checklists, and manuals from other sources and sharing them amongst members of the organization. In the early days of the process safety guerrilla organization, the "Fair Use" provision of the copyright law permits the organization to make a small number of copies from purchased books to share with a few colleagues. Usually, equations may be rendered into spreadsheets without asking permission from the publisher, but special solution methods may not. In order to legally reproduce published materials such as nomographs and other figures, tables, checklists, and text, at a larger scale, permission should be obtained from the publisher. Generally, the fees which are assessed are quite low compared to the effort involved in developing similar material from scratch, and these fees can be shared amongst plants as in the case of software.

Gaining Strength

As the process safety guerrilla organization gains strength, it will be able to take a more authoritative stand when advocating engineering designs, improvement projects, and corporate programs. As reported, this begins to happen naturally when several process safety guerrillas are seeded into the plant leadership team and corporate leadership teams, with one guerrilla supporting the points of view of the others.

One respondent reported that even before process safety guerrilla groups become legitimized in a top-down system, other engineers may begin to recognize them as a necessary stakeholder from which to obtain buy-in for new activities, much in the same way it may be necessary to obtain the support of a local gang before starting a business on their turf. When this begins to happen, the process safety guerrillas are ready to make their big move.

Gaining Control

Like a traditional guerrilla operation, the ultimate objective of the process safety guerrilla operation is the orderly transition to a stable government which operates under the principles espoused by the guerrillas. This may naturally occur by maintaining and growing the guerrilla activity for long enough that guerrillas become promoted to senior positions.

The process may be accelerated by converting senior managers to the cause. However take caution to avoid being too aggressive. One respondent, impatient for action, reported that he requested an audience with the CEO of the company to discuss a serious condition in a manufacturing process he had identified, and to offer his recommendations to address this condition. The CEO listened respectfully, agreeing that the risk to the company was unacceptable. However, instead of following the recommendation to improve the process and establish a strong process safety program, the CEO instead decided to sell the business to another company. The company's risk was reduced, but the risk was only transferred to another party.

Another way to gain control, given sufficient strength of the guerrilla organization, is to capitalize on serious accidents and near misses. In one reported case, an executive was ready to exit a business after a major accident. The guerrillas met with the executive to explain that the accident was instead a warning about other potential accidents in the company, and therefore exiting the business would not solve the problem. Instead, they recommended that the company learn from the accident so that they could avoid other accidents and gain a risk-reduction advantage over their competition.

The important thing to recognize is that the transition from guerrilla organization to corporate sanctioned program must ultimately happen according to good leadership and corporate business practices. When successful, the guerrilla organization must give up its guerrilla identity in exchange for top-down business support. The good news is that CCPS literature (see references) is full of roadmaps and other tools for selling process safety management programs to senior management.

PROCESS SAFETY GUERRILLA TACTICS

Gang Activity

If you've worked in industry in the last 25 years, at some point you've worked on at least one team. A successful team has a corporate champion, a defined and supported leader, clear and measurable goals, clear responsibilities, and a defined end result. As described above, a process safety guerrilla group cannot be organized as a team; more accurately, they organize as gangs, i.e. loosely organized groups with a common ideology and social cohesion. Process safety guerrilla gangs may or not follow the model of urban gangs to utilize common modes of dress, slang, or symbols. However, these approaches may help build cohesion and alignment between gang members.

For example, one guerrilla gang took to clipping their hair close at a time when hair styles were longer. While hardly a statement on par with the low slung jeans of urban gangs or Mao's blue gabardine jackets, the hairstyle created an additional common bond between the gang members, and allowed them to know immediately when new members joined.

Successful process safety guerrilla efforts may give up the "colors" they displayed while being a guerrilla, however, creating new colors for the new corporate effort may make sense. There are plenty of "Think Safety" mugs on employees' desks to bear out this point.

Theft or Diversion of Resources

By definition, process safety guerrilla efforts do not receive corporate funding or support. So process safety guerrilla gangs must gain the needed resources by stealing them. This paper does not advocate theft of money or property, but instead suggest the following examples of process safety guerrilla theft:

- Stealing ideas shamelessly: Whether you are hearing this presentation live or reading it in the conference proceedings, you're doing this right now. Attending conferences, webinars, and presentations at AIChE local section meetings is a good way for process safety guerrillas to pick up technologies and techniques to support their effort.
- Leeching on other corporate programs: The rigor of a good process safety management system is akin to the rigor of a quality management system. If the opportunity to implement a quality management system or six-sigma effort is presented, process safety guerrillas should jump at the chance. Such efforts support similar principles to process safety such as prevention and well-designed chemical and business processes, so other participants should be receptive to these concepts. As one respondent reported, quality is driven by the customer. Since the customer also required a reliable supply, the guerrilla gang was able to include reliability management in the quality management system, and the company's process safety record benefited.

Sabotage, Arson, and Assassination

Guerrilla tactics differ from standard military tactics in that the standard military approach tends toward direct attack at the enemy's strength, while guerrillas tend to, in Mao's words, "Retreat from strength, attack weakness." Without top-down support, process safety guerrillas stand little chance of success with direct implementation of an improved process safety program. So, while organizing and gaining strength, process safety guerrillas can try to put stumbling blocks in the way of wrong-thinking programs; in a positive, constructive way, of course.

More than one process safety guerrilla listens closely at department meetings while colleagues discuss new processes under development, and then later engages colleagues who appear to be off-track in terms of hazard identification and process safety design to sabotage the poor design with constructive solutions. When doing this, it's recommended to avoid claiming credit for the improvement, as this makes colleagues less willing to listen the next time and instead slowly turns them to the cause.

Guerrillas also commonly reported being given a lab process to scale-up "as-is." These guerrillas then had to delicately send the process up in flames through showing that the powder being handled is toxic to handle or has an unusually low Minimum Ignition Energy (MIE), or assassinate the process through calorimetry tests which show the reaction temperature bordering on the Self-Accelerating Decomposition Temperature (SADT).

Official Corruption

Process safety guerrillas can make significant progress towards their aims by identifying managers senior to them in the organization that they can corrupt to the cause. A guerrilla seeking to do this should choose their candidate carefully by observing for the telltale signs of proactive, holistic thinking. Such a candidate manager can then be approached, as one person did, by establishing a casual relationship with them in a neutral setting, and slowly over time begin engaging them in discussions about the operations or process. Once a comfortable relationship is established, the person noted an on-going near-miss situation and suggested a solution. Sensing the manager's receptivity to the suggestion, she observed that numerous such situations existed throughout the unit, and the manager was ultimately corrupted to the cause.

Corruption of a Minor

Unfortunately, not many universities introduce chemical engineers to the basics of hazard identification and safe design. Process safety guerrillas should therefore not limit themselves only to their plant and company, but should also seek out university students to corrupt. Every US University has an AIChE student chapter, each needing to program eight or nine monthly meetings and eager to accept offers to speak. Almost every other University in the world also has a chemical engineering society chapter with similar needs. Likewise, many design course professors are happy to welcome guest speakers from industry.

The good news is that today's students are very receptive to learning about process safety and environmental protection, and many are eager to apply their chemical engineering training towards these aims. Yes, process safety guerrilla actions can take many years, and it's good to keep attracting young recruits.

YOUR LITTLE RED BOOK

The author understands that readers expect him to mention the CCPS books "Guidelines for Risk-Based Process Safety (5)," "Guidelines for Hazard Evaluation Procedures (6)," and "Guidelines for Mechanical Integrity Systems (7)," all available at www.Wiley.com/go/aiche. However, these are all big red books, and like Mao's guerrillas, process safety guerrillas need a little red book. Therefore, a copy of this paper, in "little red book" form can be obtained on the CCPS website. Readers who would like to suggest enhancements or illustrative examples to improve the process safety guerrilla's little red book are invited to send these to the author for inclusion in periodic updates.

Figure 1 is offered as a roadmap to future process safety guerrillas to guide them on their journey.

In conclusion, a strongly supported top-down process safety culture and program remains the ideal way to give the company the best possible chance to implement a sound process safety program that prevents incidents. However, individuals who do not work for such companies still have a way to address process safety concerns and ultimately establish the desired top-down state over time.

ACKNOWLEDGEMENT

The author gratefully acknowledges the contributions to this article of the individuals who shared examples of their process safety guerrilla activities, both for their contribution to this article and for their contribution to process safety.

REFERENCES

1. CCPS, "Guidelines for Technical Management of Process Safety,"AIChE, 1989
2. Mao, Zhedong, "Principles of Guerrilla Warfare, China,1937
3. Lenin, Vladamir, "Guerrilla Warfare", *Proletary*, No. 5, Russia, 1906
4. Kletz, Trevor, "Training by Discussion," Trans IChemE, Part D, 2006, Education for Chemical Engineers, 1: 55–59
5. CCPS, "Guidelines for Risk-Based Process Safety," AIChE/Wiley, 2007
6. CCPS, "Guidelines for Hazard Evaluation Procedures," AIChE/Wiley, 2008
7. CCPS, "Guidelines for Mechanical Integrity Systems", AIChE/Wiley, 2006

Figure 1. Roadmap of a Process Safety Guerrilla Campaign

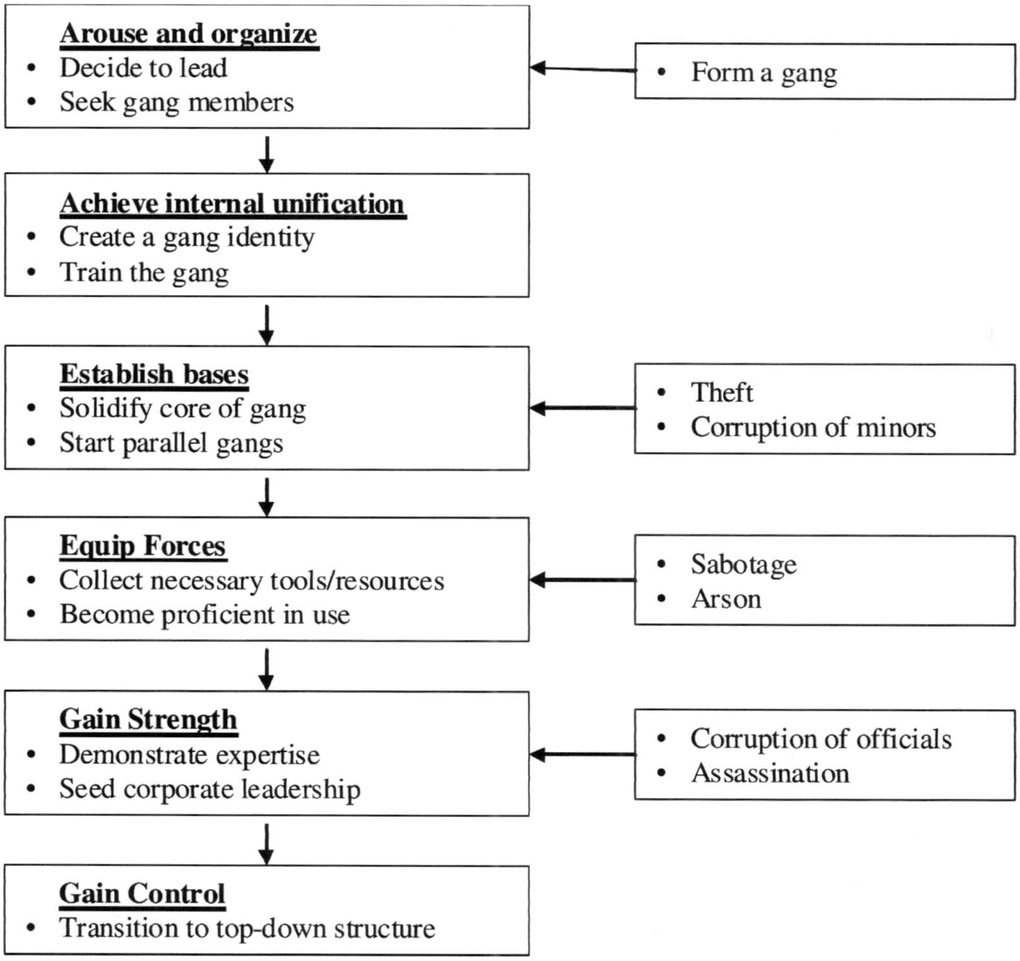

Using Playbooks – A Unique Strategy in Technical Oversight of Highly Hazardous Operations

James Gary, Shaw E&I, Inc., Newport Chemical Depot, Newport, IN
Jeffrey Brubaker, U.S. Army, Chemical Materials Agency, Newport, IN
Scott Haraburda, U.S. Army, Crane Army Ammunition Activity, Crane, IN
Michele Summers, Purdue University, College of Technology, Lafayette, IN
Lee Gary, Strategic Management Services, New Orleans, LA

ABSTRACT

The US Army completed VX nerve agent destruction in August 2008 at the chemical weapons disposal facility in Newport, Indiana. Since VX is a very toxic chemical warfare agent, operational errors, mishaps and other unplanned events during handling and destruction are totally unacceptable.

A systems contractor operated the facility; but, the Army had responsibility to ensure operations were safely conducted per laws, regulations and applicable permit conditions. To protect site workers and the surrounding citizenry, a new and unique approach for defining and articulating "Technical Oversight" of operating preparedness, emergency response, and recovery was created. Technical Oversight was the duty of independent representatives, who were "certified" by the US Army as "Government Shift Representatives (GSRs)," and served as the on- scene "eyes and ears" of the Government Site Project Manager (SPM).The GSR Certification Program utilized a rigorous six-phase process to prepare the GSRs for assignment, including:

1. **Overview Training** on how to conduct oversight duties
2. **Technical Process Training** which was the same training received by all plant operators, mechanics, and engineers
3. **Required Readings** of various plans, procedures, and reference documents
4. **Critical Systems Demonstration** to communicate understanding of equipment locations and functions
5. **Oversight Effectiveness Training** on Conduct of Operations topics
6. **Oral Board Examinations** to confirm readiness for oversight duties.

The key phase, Oversight Effectiveness Training, was based on playbooks, which was the unique medium utilized to ensure uniformity in awareness and understanding between all GSRs. Playbook guidelines were individually prepared by the GSRs on various Conduct of Operations topics, and to ensure consistency, the GSRs participated in Roundtable Sessions focused on discussing the details of each guideline. As with a football playbook, having the entire team learn the plays(guidelines), practice them, and recognize them when called (or directed), maximizes the chances that all team members will perform the right way and exactly the same way every time.

The Newport site utilized this disciplined approach to develop and execute an effective oversight program during the demonstrated safe destruction of 1269 tons of VX nerve agent. With these playbook guidelines serving as the cornerstone of the certification process, a cadre of GSRs possessing the requisite skills and knowledge were available for successfully executing technical oversight of the VX destruction. A similar approach can be used by other projects and programs needing rigorous, consistent oversight.

SUMMARY

The Chemical Stockpile Disposal Program (CSDP) is a U.S. Army program implemented to destroy the nation's inventory of Chemical Warfare Agents. One of the stockpile locations was Newport, Indiana where VX nerve agent had been stored since the 1960's until successfully destroyed in August 2008. The Newport Chemical Agent Disposal Facility (NECDF) was designed and built to neutralize VX instead of using the Army's baseline technology of incineration. The incineration technology was utilized at several other stockpile locations and had provided an extensive cadre of experienced oversight staff and operational knowledge developed over 15 to 20 years. The NECDF project utilized a different destruction technology; thus, it was necessary to develop new competencies within the Newport Government team responsible for the oversight of the systems contractor tasked with destruction of the chemical weapons. The Government team used a formal qualification and certification process to develop and test these competencies. Utilizing the processes described in later sections of this paper, NECDF certified 16 GSRs over a period of six months to support initiation of VX stockpile destruction in May 2005.

INTRODUCTION

Many extremely hazardous operations are executed every day with the common practice of owners having oversight programs to ensure they are being performed safely and within expected parameter ranges. For oversight to be effective, those individuals performing oversight must be highly trained and consistent in the discharge of their duties. It is the responsibility of the owners to establish a process whereby desired technical competence and uniformity of oversight are achieved. This paper outlines a proven process for delivering the desired results for oversight programs of highly hazardous operations.

Highly hazardous operations fall into several categories, including chemical plants manufacturing reactive or explosive compounds, industrial construction involving huge and high lifts, transportation of flammable or acutely toxic materials, medical practices associated with potentially life threatening procedures, and power plants utilizing nuclear reactors to produce energy. All of these activities use procedures and practices requiring highly trained and skilled professionals and technicians. Yet, they are subject to errors or mistakes that could be extremely costly or result in termination of the activities or even the organization performing them. Most owners or senior executives responsible for such operations, therefore, establish independent oversight teams to monitor and report on the performance of the executing organization. It is extremely important that these oversight teams have technical skills at least equal to or preferably greater than the organizations monitored so that they will be able to quickly detect deviations from desired performance. It is also critical that these oversight activities are consistent from individual to individual on the oversight team, which typically could include anywhere from as few as four to over twenty individuals depending on the activity and organization being oversighted. These requirements of technical skills and uniformity are important because inconsistent oversight can result in damaging the organization being monitored more than if there was no oversight at all.

Thus, it is critical that the oversight team is prepared by ensuring that all its members are technically trained and proficient, and perform in a consistent manner at all times. There should be no surprises from the oversight team that cause the organization being monitored to react inappropriately, ineffectively, or unsafely. To ensure such desired results from the oversight team, they must be trained and certified as professional oversight personnel with the skills to interchange in all aspects. The goal is to develop an oversight team member with the best of all the skills from all the other team members and certify those individuals so owners and senior management can be assured of consistent and high quality oversight of their operations. This paper focuses upon the details of the process employed at the NECDF to achieve these desired goals and contribute to the outstanding performance in destruction of the VX chemical agent.

ESTABLISHING THE NEED FOR OVERSIGHT CERTIFICATION PROCESS

The most effective oversight process is one that is targeted to deliver a team of technically competent and interchangeable individuals. That is, all members of the team must demonstrate technical knowledge and proficiency that meet minimum standards. They must also be capable of being switched from one assignment or area to another with no significant difference observed by the oversighted organization. These criteria are important because oversight team members must be able to engage the oversighted organization on sound technical bases that are recognized and accepted quickly and easily with minimal disruption to the executing organization. Also, each oversight team member must comply with the agreed rules and processes of the oversighed organization without freelancing and without developing new or different standards. Failing to meet these criteria can result in upsetting the oversighted organization to the point that it must pause or cease operations, or worse, proceed in an unsafe or environmentally unacceptable manner.

To achieve the criteria discussed above requires a robust process that trains and certifies oversight personnel consistently and generally on a fast-track basis. Like most processes, given enough time and resources, nearly any objective can be achieved. However, extended schedules can become quite costly and unacceptable to owners. The most desirable process is one that is technically successful, but also capable of being expedited if necessary. An expedited certification process is one that is very well defined and capable of being broken down into manageable components or phases that can be meshed and overlapped effectively. For example, while conducting classroom training, some amount of field hands-on training can also be executed, allowing certification candidates to acquire advanced technical knowledge in the classroom while concurrently putting it in practice in the field. In the following sections, the structure of the certification process utilized at the NECDF and deemed Best Practice by the U.S. Army (1) is presented in detail.

STRUCTURE OF THE CERTIFICATION PROCESS

The certification process used for the NECDF project was broken down into three areas, as follows:

1. Qualification. Qualification activities dealt with improving knowledge and understanding of technical and background information. It dealt with the theory of various principles and processes upon which the operation was based. Generally, qualification involved training in a classroom setting and exposing the GSR to academic type functions, such as reading documents, class lectures and discussions, studying history and lessons learned from similar operations, and satisfactorily completing tests and written examinations.

2. Certification. Certification involved GSRs demonstrating the knowledge and principles obtained during qualification by applying them to the process and organization being oversighted. While qualification normally occurred in a classroom setting, certification activities were performed "in the field," or where the activities and operations were to take place. The GSRs normally did not operate equipment or execute actual operations tasks; but for certification, they were asked to physically locate various items or equipment and describe their operation to an instructor or trainer. In this way, GSRs demonstrated that they

completely understood what the system contractor should be doing when they observed a particular operation being performed. A simple analogy for certification is that it is like taking a driver's road test under the observation of an instructor, after having studied for and successfully completed the written exam.

3. Authorization. Authorization was the final area addressed, after the GSRs successfully completed qualification and certification requirements. This area involved situational testing of the GSR's knowledge to determine if they were capable of applying the skills acquired from the preceding two areas. Finally, authorization involved the Government Site Project Manager signing off that he authorized the GSR to assume the role of a certified oversight team member.

QUALIFICATION AREA

Qualification was the first major area addressed and was composed of the following three phases: Oversight Training, Technical Process Training, and Required Reading.

1. Oversight Training. In Oversight Training, the GSR was initially introduced to the process to be utilized for conducting oversight. This phase provided definitions and background on what oversight was, in general, and how it would be executed on the NECDF project specifically. This training utilized the oversight plan and procedures that had been developed based on the government's specific desires and represented what was considered to be the detailed objectives of the oversight program being instituted. This phase was the crucial "first step" taken to begin forming the GSRs into a team that would be technically consistent and interchangeable. At the completion of this phase, each GSR had obtained a good appreciation of oversight and the processes, protocols, and procedures that were to be utilized to accomplish the goals established by the government. Many highly hazardous operations involve activities and protocols routinely encountered in typical chemical or other industrial plant settings, and they have adopted the U.S. Department of Energy (DOE) Conduct of Operations requirements (2) as Best Practice for executing those activities. These DOE requirements, which were developed for "improved quality and uniformity of operations," were also selected to serve as the basis for

the following NECDF Conduct of Operations Focus Areas Checklist:

- Operations Organization & Administration
- Shift Routines & Operating Practices
- Control Area Activities
- Communications
- On-Shift Training
- Investigation of Abnormal Events
- Notifications
- Control of Equipment and System Status
- Lockouts and Tagouts
- Independent Verification
- Logkeeping
- Operations Turnover
- Operations Aspects of Facility Chemistry and Unique Processes
- Required reading
- Timely Orders to Operators
- Operations Procedures
- Operator Aid Postings
- Equipment and Piping Labeling

This Checklist was introduced to GSRs in this training phase and subsequently utilized throughout the NECDF project as the primary focus for concentrating oversight activities. Additionally, case studies from other projects or operations were reviewed for lessons learned that could be applied to the NECDF activities to be oversighted. As an example, GSRs reviewed the events associated with the Union Carbide catastrophe that occurred in Bhopal, India in 1984 (3,4) and resulted in the death of approximately 20,000 people when the deadly gas methyl isocyanate was released. This was a good example of where successful oversight might have prevented the event, and the NECDF GSRs discussed lessons learned from Bhopal that might be applied to NECDF oversight. Supplemental cases from the U.S. Chemical Safety and Hazard Investigation Board (5) were also selected and reviewed for applicability of lessons learned to the NECDF oversight program.

2. Technical Process Training. Following introduction to the qualification area via the Oversight Training phase, GSRs then underwent rigorous process and procedures orientation as part of the Technical Process Training. In this phase, GSRs participated in the same classroom training that was being presented to engineers, operators and technicians of the system contractor to ensure they obtained the same degree of technical knowledge as the actual operating and maintenance staffs. This included studying the design of the facilities and equipment, understanding the detailed steps involved in all processes, learning the sequence and justification for all steps in procedures to be utilized, and being introduced to emergency procedures for handling incidents and accidents. Detailed reviews of vendor and supplier manuals were also performed to ensure that component items were understood as well as the aggregate facility. Regular testing of GSRs throughout the process and procedure training phase was employed to ensure understanding of all elements covered. At the end of this

phase, candidates were as knowledgeable in the details of operating and maintaining the facility and responding to emergency situations as were the systems contractor's engineers, operators and technicians.

3. Required Reading. The final phase of the qualification area was Required Reading. This was essentially a self-study program in which GSRs were assigned a lengthy list of documents that had to be read. To ensure that each GSR read the documents, they had to provide documentation that all assigned readings had been completed. There were no formal meetings or classes covering these readings; but, familiarity with their content was important to successful completion of other phases that will be addressed later in this paper. The readings were usually lengthy documents of a general rather than specific nature. In document hierarchy, plans were developed first. Then, procedures were prepared providing more specific details to the topics covered in the plans.

CERTIFICATION AREA

The second major area pursued was Certification, composed of two phases which were: Critical Systems Demonstration and Oversight Effectiveness Training.

1. Critical Systems Demonstration. Since certification focused on demonstrating that classroom knowledge acquired during qualification could be applied to actual field situations, this phase required GSRs to demonstrate their familiarity with facility operations through a "walk and talk" activity. This was necessary to ensure that GSRs did, in fact, know how the process functioned and recognized when members of the system contractor organization were carrying out their duties properly. During demonstrations at NECDF, under the observation of a subject matter expert certified to operate the facility, a GSR was asked to physically locate key systems and specific pieces of operating or safety equipment and verbally describe its major components. Then, they had to answer questions concerning designated function and performance details of the system or equipment. For example, for a particular pump, this would include responding to performance questions, such as stating the pump capacity and discharge pressure, as well as other technical and operating characteristics of the pump. While not asked to actually operate or manipulate equipment, the GSR had to demonstrate knowledge of how to perform the required functions. Additionally, the GSR had to demonstrate the capability of observing someone else operating the equipment to determine whether actions were being done safely and properly. Completion of this phase satisfied one of the two requirements that the government had for a successful oversight team - that the oversight personnel are highly trained and technically proficient.

2. Oversight Effectiveness Training. This phase satisfied the second government requirement for shift-to-shift consistency in execution among GSRs (to the point of interchangeability). Possessing this characteristic was important to avoid providing inconsistent or even conflicting feedback to the system contractor, which might result in inefficient, or even much worse, unsafe operations. To achieve this consistency, GSRs participated in a series of roundtable training and discussion sessions that focused on the details of the Conduct of Operations topics previously introduced during the Oversight Training phase. Prior to a session, assigned GSRs initially developed draft guidelines for a particular Conduct of Operations topic. The draft guideline was then presented to all GSRs at a roundtable session. After it was thoroughly discussed, it was then placed into a reference book called the "playbook." This followed the concept of a football playbook which documented what each player will do when a given play is called. As with a football playbook, having the entire team learn the plays (guidelines), practice them, and recognize them when called (or directed), maximizes the chances that all team members will perform the right way and exactly the same way every time. After each playbook guideline was presented to the rest of the GSR Team for discussion and training purposes, it was either accepted or revised and presented again to the group later for final acceptance. This process was continued until all assigned topics were adopted by consensus and accepted by all GSRs. Once a playbook guideline was approved, the GSRs were committed to consistently follow it in the execution of their oversight duties. More details will be provided on the preparation and use of these playbooks later in this paper.

AUTHORIZATION AREA

The final major area, Authorization, was composed of two phases as follows: the Oral Board and Management Sign-Off.

1. Oral Board. After certification activities were completed, GSRs possessed all the theoretical knowledge (qualification) and applied knowledge (certification) necessary to conduct effective oversight of the system contractor. What remained was to have GSRs sit for an Oral Board examination that covered all of the previously described phases of training, reading and demonstrations. The Oral Board was composed of senior owner representatives and selected experienced personnel, as well as experts from outside sources, such as other similar operating sites. Prior to the Oral Board, review questions were distributed to all GSRs for use in preparation. In addition to specific technical questions, the board prepared situational questions designed to explore the GSR's ability to

apply their technical knowledge to specific situations. These types of questions were important to predict how a GSR might react to conditions that were not routine or easily anticipated. For example, the board might ask how GSRs should react to observing a system contractor employee preparing to make a transfer of highly hazardous material from one tank to another, but appearing to be intoxicated or physically impaired. Prior to sitting for Oral Boards, GSRs met as a team and developed, then reviewed and discussed, acceptable answers to these preparation questions. This activity represented a good team building activity that helped to further increase knowledge and solidify the GSRs as a true team. Based upon a few hours of individual examination and discussion, the Oral Board then determined if a GSR should be recommended to senior management for final approval and certification as an oversight team member. If unsure, the board recommended further study and subsequent Oral Board reevaluation.

2. Management Sign-Off. Once a favorable recommendation for a GSR was received from the Oral Board, they met with the Government Site Project Manager. The purpose of this Management Sign-Off session was to allow for final discussion between the manager and the GSR to ensure there was good mutual understanding of the government's vision, strategies and goals. Additionally, this meeting provided an opportunity for discussion of any final questions or concerns that the GSR might have. A review of the GSR's performance in each of the completed phases of the certification process was also reviewed and discussed. If, at the end of this session, the Government Site Project Manager and GSR agreed to proceed with final authorization, then the manager signed off on the official approval record, resulting in the GSR becoming a certified oversight team member.

USING PLAYBOOKS AS THE KEY TO SUCCESS

For the most part, the NECDF oversight certification process was similar to other structured processes for training and certifying individuals for specialized assignments. The unique element, however, that set it apart from others was the playbook feature. While most phases of training and certification programs were passive for the participants, playbooks provided a mechanism for GSRs to assume a proactive leadership role. Unlike other phases where the training materials were previously prepared by engineering or procedure technical writers and simply passed on for use, Playbook Guidelines were developed from scratch and allowed GSRs the opportunity to conduct research and develop the guideline to their style.

Fortunately, some of the individuals who were selected to serve as GSRs did have past plant experience in actual shift operating organizations. Those who did not were, however, familiar with industrial plant activities in general because they had experience from assignments in engineering, construction or startup of plant facilities. One of the strengths of the playbook program that was instituted for NECDF was that individuals with general plant experience could be "re-tooled" quickly to serve as a GSR. This conversion drew heavily upon the individual's general industrial experience and familiarity with NECDF specifically, but also utilized those staff members experienced in actual operations to serve as mentors and coaches. Thus, the GSR Team formed at NECDF had a broad range of experience and every member learned from and drew upon the best practices of each other. Since using playbooks provided a systematic and expedited process for training individuals for oversight roles, NECDF was able to retain staff from the engineering, construction and startup phases of the project and not have to recruit new individuals that would have no familiarity with the NECDF site and process. Thus, the playbook program was beneficial for both the individuals already on the NECDF project and the government leadership responsible for staffing the oversight activity. Since many of the GSRs may not have had past experience in preparing similar documents, the GSRs were provided general directions for developing playbook guidelines and were coached on how to proceed based on the following information:

- Research of project documents
- Review of handouts from other oversight training phases
- Discussions with knowledgeable individuals at the site
- Discussions/visits with key personnel at other similar sites
- His/her own personal experience
- Any other source deemed appropriate by the preparer

Using the above information, GSRs researched all project documents, such as plans, procedures, reports, permits, engineering studies, and publications to locate any background information that may be applicable to the Playbook Guideline they had been assigned to prepare. Also, they utilized all handouts and instructional materials from other training they may have completed, such as Technical Process Training or Required Reading, for material that would add to their final product. It should be noted that GSRs were assigned topics that they did not necessarily fully understand, so it was a learning experience for them and they had to dig deeper and research harder than someone who was already a Subject Matter Expert (SME) on the topic. After completing the research activity, the assigned GSR was encouraged to contact people knowledgeable on the topic both currently at the site and also at other similar sites to discuss it with them. There were several other sites that had completed or were currently conducting chemical demilitarization activities similar to what NECDF would be doing. Therefore, the GSRs obtained first-hand SME input for their guidelines and furthered their networking capabilities for the future. Lastly, GSRs were encouraged to bring their own experiences to bear on the preparation of their guidelines. While most had never worked in a chemical demilitarization facility before, they did have similar experiences at highly hazardous chemical plant operations, nuclear facilities, or previous military experience, and they were challenged to appropriately incorporate it into their NECDF assignments. For example, Lockout-Tagout (LOTO) procedures at chemical and industrial plants were essentially similar to those used at NECDF, and Shift Turnovers and Logkeeping utilized the same principles at nuclear facilities and NECDF. Lastly, GSRs preparing Playbook Guidelines were asked to think about other sources of information they could use to add to the quality of their guideline. This included using information from attending a conference or seminar on plant operating protocols, seeing a movie on industrial practices that lead to catastrophes, or simply having conversations with highly experienced plant operating personnel about specific events encountered. Each Playbook Guideline was prepared following a prescribed template that included all the key elements to be addressed. This Guideline was composed of three sections: Background Information, Desktop Instructions, and Guideline Details.

1. Background Information. This provided information on the topic covered in the guideline, who prepared it, configuration control information for dates prepared, discussed and revised, and lastly, references utilized. Also, this satisfied the requirement for information needed to support and maintain the document as up-to-date.
2. Desktop Instructions. The Desktop Instructions were intended to be a review and reference source for GSRs in the field when they needed a quick refresher on the topic.
3. Guideline Details. The Guideline Details could be used to bring the GSR up to date and current on the topic, when needed.

GSRs utilized the above listed and discussed guidance to prepare their initial drafts which they presented to the rest of the group. The emphasis here was on "drafts" because the initial presentation was based on what the preparer had individually developed. The final and approved guideline represented the consensus input and opinions of the entire group. The roundtable process for moving from draft to final represented the central strategy of this activity. This was also a way to build teamwork and move from individual products to team products. In this way, a true consensus was achieved, resulting in all GSRs buying into the guideline as being their own and having their commitment to support and follow it consistently. This was very important because consistency of oversight activities was a crucial characteristic for the NECDF oversight program. The systems contractor had to understand that issues would be dealt with the same way every time, so they would not have to worry about a possible change in oversight execution that could cause them to lose focus on their operation.

The Playbook Guidelines were finally accepted by all GSRs and became part of their resource documents often referred to later while performing their oversight duties. If the GSR needed a quick refresher or just wanted to review a topic in more detail, the Playbook Guideline was the first document used.

MAINTAINING THE CERTIFICATION

Once certified, NECDF GSRs were assigned to various areas and work schedules prior to performing their oversight duties. While they were part of a larger oversight team, each GSR generally worked alone and had minimal contact with the other team members. To maintain technical expertise, several techniques were utilized at NECDF: including use of reports, meetings, required readings, lessons learned, procedures and protocols, a Standing Order process, government feedback, refresher training, and annual recertification.

- Preparing and sharing daily reports. GSRs individually prepared and issued daily reports covering their observations and discussions with the system contractor. Each report was distributed to the government managers, to the system contractor management team, as well as all to the other GSR oversight team members. Thus, GSRs were kept abreast of issues that had arisen and how they were handled by another GSR. This allowed them the capability to reflect upon how they would have handled that particular situation. If necessary, GSR team members asked questions of the individual filing the report for additional information and details to ensure they completely understood the situation. Finally, these reports were maintained in a database for easy retrieval and reference later as needed.

- Conducting routine weekly meetings. On a regular basis, usually weekly, a meeting of all available GSRs was held to review and discuss activities from the previous week. Primary attention was given to issues or concerns that appeared in the daily reports so that the GSR team could focus upon whether they had been technically proficient and consistent. Tremendous effort was placed on not being critical of individual styles or techniques but whether the actions taken were appropriate. Team leaders met privately with individual team members to coach them on style or techniques; but, open team discussion focused on technical proficiency and consistency. The team also looked for recurring themes that may indicate a systemic problem with system contractor's operations or organization; and, if determined to be present or emerging, government senior management was made aware of that determination for follow-up action. Finally, the GSR team usually discussed particular areas or topics that should be watched more closely in the upcoming week.

- Completing assigned ongoing Required Readings. Just as required readings had been a part of the qualification area during the certification process, all GSRs continued to participate in a required readings program after performing oversight. The purpose of this activity was for them to stay current on new issues or information that could affect their oversight duties. The readings were from technical journals or periodicals, revised procedures from the systems contractor, and general reports from government oversight management. This program kept all GSR team members informed to the same level even if they did not have daily or routine contact with other team members or management.

- Reviewing applicable Lessons Learned from other similar oversight organizations. While required readings tended to keep GSRs informed of general information about oversight activities or proceedings within the systems contractor organization, reviewing lessons learned from other similar organizations provided additional specific and applicable knowledge. Lessons learned were usually issued when an organization obtained information about a past event or incident that could be helpful in preventing reoccurrence under similar conditions. Reviewing these lessons learned and determining how they could be applied to another organization or activity was useful in that they provided scenarios for consideration. GSRs could study the possibility of a similar event or incident occurring in the NECDF systems contractor's organization oversighted.

- Periodically reviewing/updating the procedures and protocols being used for conduct of oversight. Like the systems contractor, the GSR oversight team conducted regular reviews of their procedures and protocols and ensured they were current and technically correct. This provided another opportunity for keeping the GSR current while confirming that the procedures resulted in consistent performance by all team members. These periodic reviews also generated questions that were added to the regular weekly meeting agenda for further discussion.

- Utilizing a Standing Orders program/process to ensure GSRs provided extra focus where needed. Standing Orders were generally used to supplement the standard operating procedures. The NECDF project leadership used Standing Orders to direct the actions of on-shift GSRs when more focus was needed or specific observations were required in a particular area or activity. Knowing how to execute Standing Orders and document observations made in support of them was an important skill that had to be maintained by all GSRs. Standing Orders were reviewed and discussed at weekly meetings.

- Reviewing feedback from the government on oversight being performed on the systems contractor. Feedback should be welcomed and evaluated for possible use in continuous improvement. Oversight organizations should be subject to critique and evaluation, but may not receive it directly. For NECDF, the most likely source of such feedback was the government management staff. The systems contractor provided information to them that they would not share directly with the GSRs. This was particularly true if the feedback applied to a specific individual (as opposed to the entire team). When such feedback was received at NECDF, it was evaluated and utilized to improve performance, generally in an open and positive manner. When the feedback referred to a particular GSR, it typically resulted in additional coaching or mentoring for that individual, to make sure their certification and continuation in the program was not jeopardized.

- Receiving new or refresher training by attending Playbook sessions. The GSR team recognized the need for new or refresher training in order to maintain technical proficiency and consistency. The best process for conducting such training was preparation and delivery of a playbook session for all GSRs. Since the GSR oversight team members had become familiar and comfortable with playbook techniques during certification, it was a readily accepted method of training for all. These playbook sessions quickly added the topic(s) covered to the body of knowledge needed to maintain certification for all existing and new GSR personnel who later joined the team.

- Annual recertification after review of individual performance and validation of competency. Certification of NECDF GSRs was for a specific period of time, normally one year. Therefore, annual recertification was necessary. The following items were reviewed annually to determine if recertification should be granted:
  - Accurate and timely submission of all daily reports.
  - Good attendance at all required weekly meetings.
  - Maintaining current status on all required readings.
  - Evidence of timely review of all issued lessons learned.
  - Contribution to periodic review of procedures and protocol.
  - No unresolved issues concerning negative feedback from the government staff or systems contractor personnel.
  - Attendance at all required on-going playbook sessions.
  - Periodic presentations of playbooks at routine meetings.
  - Favorable discussion with appropriate government manager responsible for signing off on annual recertification.

Once all required recertification criteria were met, individual GSRs were recertified for another year. If any situations or conditions arose concerning the appropriateness of an individual certification before the next anniversary due date for review, an adhoc recertification evaluation was initiated. If such an event did occur, the individual GSR was usually temporarily removed from their oversight duties until the recertification process was completed. This step was essential because the risk of inappropriate oversight by a GSR was simply too risky a situation for the government to tolerate at NECDF.

SUMMARY

After completing construction and systemization testing of the NECDF, operations to neutralize 1269 tons of VX Nerve Agent began in May 2005 and were completed in August 2008. During that time period, the facility established outstanding safety, health, and environmental performance and recorded the lowest OSHA Recordable Incident Rate in the family of six Chemical Demilitarization processing sites for 22 of those months. Since the end of operations and into closure of the facility, NECDF has maintained the following excellent performance (6):

- OSHA Recordable Incident Rate of 0.36 (12 month rolling average).
- No OSHA Lost Workday Case for 12 months.
- No environmental enforcement actions against the facility.
- No events where chemical agent migrated outside of engineering controls.

To achieve this outstanding performance required the commitment of everyone, including the site government team, the systems contractor, several teaming subcontractors, local community and regulatory oversight agencies, and their corporate organizations. While all of them played important roles in the success of the NECDF project, the government field office staff had the lead responsibility for the project and relied heavily upon its GSR oversight team on a 24/7 basis throughout the agent processing period. The GSRs provided accurate and consistent feedback to the government on the overall technical performance of the project; and thus, the government leadership team was able to effectively oversight the systems contractor and emphasize the importance of lessons learned and continuous operations improvement. The GSRs executed their oversight observation and communication roles in an excellent manner because they had received thorough and complete training in their certification program, and the heart of that program was the Oversight Effectiveness Training using Playbooks.

There are many structured training and certification programs available for preparing members of teams to execute a particular function. Nevertheless, the uniqueness of the NECDF program was the use of Playbooks to establish consistency in discharging oversight activities. Through Playbooks, GSRs were able to come together initially as a group of individuals and transform themselves into a "team" that uniformly and consistently provided technically proficient feedback to the government leadership team. The NECDF project has documented the lessons learned from the safe and successful destruction of the VX Nerve Agent at the Newport facility, and will be sharing those lessons with the two remaining Chemical Demilitarization projects scheduled to operate in the United States in coming years. One of the lessons being stressed is to ensure consistent and technically proficient oversight of all activities through a certification process that uses a technique similar to the NECDF's Playbook program.

The Playbook program, process, and techniques used at the NECDF facility can be used at other highly hazardous operations. It was successful because it stressed the importance of proactive participation of all team members in the preparation of the guidance used by the entire team. Often, individuals receive direction from procedures or other directives but never really understand the need to accept and execute them uniformly and consistently. So, they independently develop their own interpretations of what the guidance means to them, and act accordingly. By using Playbooks to gain a common understanding and commitment to consistency, individuals can be transformed into uniform team members.

REFERENCES

1. Haraburda, S.S., and J.L. Gary, "Certification for Government Oversight of Manufacturing," *Defense AT & L,* Vol XXXIII, No. 4, DAU 180, pp. 46-49 (July-August 2004).

2. "Guidelines for the Conduct of Operations at DOE Facilities," U.S. Department of Energy, Order DOE 5480.19, Washington, D.C. (Oct. 2001).

3. Eckerman, I., "Chemical Industry and Public Health: Bhopal as an Example," Essay in Master of Public Health, Nordic School of Public Health, Goteborg, Sweden, 2001

4. Kalelkar, A.S., "Investigation of Large-Magnitude Incidents: Bhopal as a Case Study," Presented At The Institution of Chemical Engineers Conference On Preventing Major Chemical Accidents," London, England, May 1988.

5. "Hazard Investigation: Improving Reactive Hazard Management," U.S. Chemical Safety and Hazard Investigation Board, Report 2001-01-H, Oct 2002.

6. Maynard, Ward, "12-month Rolling RIR for Chem Demil = 0.89 World Class," e-mail dated 2/5/2009, US Army Chemical Materials Agency, Edgewood, MD.

Exhibit 1—Typical Playbook Guideline

PLAYBOOK GUIDELINE
Guideline No.: 9

Background Information

Topic:	Lock Out/Tag Out (LOTO)
Prepared By:	J.L. Gary
Date Prepared / Presented:	22 March 04/25 March 04
Date Accepted:	24 May 2004
Revision:	1 (4 June 04)
Reference Documents / Procedures:	-DOE Conduct of Operations, Order # DOE 5460.19. -Chapter: IX, NECDF Conduct of Operations Plan (CDRL-A-022): a) Rev 1(21 Jan 2004) – Section Number 2.2.2, Safety Practices. b) Rev 1 (21 Jan 2004)—Section No. 4.8.1., Periodic Verification. -Procedure PP-SA-05. -OPFI (Operating Field Instruction) - ADM-000-004,Lockout/Tagout.

Desktop Instructions

- Periodically audit a LOTO Log with Plant Shift Manager (PSM) and Shift Safety Rep to determine if Lock out/Tag out OPFI is being properly implemented, and using the latest approved and controlled revisions.
- Investigate and confirm that only ***authorized employees*** are performing lock out/tag out activities or completing maintenance work on equipment that is locked/tagged.
- Check for Concurrent Verifications being used when required per OPFI.
- Observe for performance of "test-by-touch" (or test run to verify zero energy practices) by both Operations and Maintenance personnel, as required by Lock out/Tag out OPFI.

Guideline Details

LOTO programs provide methods for controlling the status of equipment via specific Locks and Tags.

The purposes of LOTO programs are to:

- -protect personnel from injury
- -protect equipment from damage
- -protect the environment from spills/releases
- -maintain operability of plant systems
- -establish and maintain integrity of physical boundaries of plant systems

during equipment operation, servicing or modification activities.

LOTO programs should provide for verification of the removal from service and restoration to service of facility equipment. This is important because improper removal or restoration could result in serious injury or equipment damage. NECDF will do ***concurrent*** but not ***independent*** verifications.

LOTO programs must contain provisions for "test-by-touch" to verify and confirm that energy sources have been successfully isolated and all pressure or process fluids have been vented or drained.

LOTO programs must meet the requirements of 29 CFR 1910.147, which are the OSHA rules for LOTO.

NECDF meets the DOE Conduct of Operations and OSHA requirements for LOTO thru implementation of OPFI-ADM-000-004, which is the Field Instruction for Lock out/Tag out. Read and become thoroughly familiar with this OPFI, so you can recognize the details of implementation in the field, such as different colored locks, lock boxes, Lock out/Tag out Logs, to mention a few. If unsure about details of this OPFI, retake the training class on LOTO or meet with Parsons trainers for specialized instructions. Also, be aware that Lock out/Tag out procedures are often revised many times during the early life of a new facility, so always check to make sure you are using the latest version of the Lock out/Tag out OPFI.

By Lock out/Tag out OPFI, the Shift Safety Rep is required to audit the LOTO program. Make arrangements with the Shift Safety Rep to include you whenever he plans to conduct such an audit and go along with him to observe and confirm compliance with the OPFI for yourself. Additionally, QA/QC may be performing separate audits, and if you become aware of such an audit, include yourself in those also.

Only ***Authorized*** Employees who are trained to lock out or tag out machines or equipment (Operations Authorized Employees) or perform work on such equipment (Authorized Employee) can do so.

Employees who are trained to operate equipment on which maintenance is being performed under LOTO or whose job requires them to work in an area where the maintenance is being performed are called ***Affected*** employees.

Lists are prepared by Training showing who are Authorized Employees for LOTO. Parsons is currently evaluating where those lists will be posted, but it will be in either the CON or Ops Trailer.

Locks are to be used in non-Chemical Exclusion Area but not used inside the Chemical Exclusion Area. Lock out/Tag out for equipment inside the Exclusion Area will be accomplished by electrical disconnects or pulling fuses outside the Exclusion Area, but be advised that this activity is still being developed and new or different information may be received at a future date.

An Impetus for Management Involvement in Process Safety

Jack Chosnek, KnowledgeOne
Houston, Texas
jc@knowledge1.net

ABSTRACT

Process Safety Management (PSM) at a facility is typically implemented to satisfy regulatory requirements because it is the minimum necessary for successful management of the facility. Ultimately managers are rated by top management on good financial performance, compliance with regulations, and an absence of safety, health and environmental incidents. Thus, lagging indicators such as the OSHA (Occupational Safety and Health Administration) log are commonly used resulting in incidents when process safety performance has not been up to par. For the most part, good PSM performance is left to the interpretation of process safety professionals in the facility and, in the absence of incidents, is not viewed as part of good overall management performance.

If practical process safety performance indicators, similar to financial indicators, were provided to management, it would give them a tool to rate good PSM programs and provide an impetus for superior performance in this area. These performance indicators should be, similar to the financial ones, relatively simple to understand and give top management a measure to which site management would be held accountable.

It is proposed that these indicators be based on activities that increase the short-term risk of the facility, such as open Management of Change (MOC) items, unresolved Process Hazard Analysis (PHA) recommendations, out-of-service critical instrumentation and unaccomplished refresher training, among others. Longer term measures such as audit findings, incident investigations, PHA completion, emergency manual upkeep, etc., would continue to be traditionally monitored. The paper will discuss and describe how these indicators would be structured, the method and frequency of their determination and the potential corrective actions that would improve performance.

INTRODUCTION

Good process safety management (PSM) requires management involvement in order to go beyond just regulatory compliance and to set a safety culture in which the principles of PSM can be properly applied. In this paper we will take a very pragmatic approach to explore the means to get management continually involved in process safety. The objective is to develop simple indicators that, analogous to financial performance indicators, will provide a continuous measure of PSM performance and an impetus for management to demonstrate good performance.

An indicator provides two functions: it lets the manager know how well he's doing and it serves as a communication tool to upper management and to the employees. The indicator will use metrics to come up with a value or values that can be communicated and easily be understood by the target audience. That is, no matter how technically correct and complex the metrics are, the distilled indicator has to be simple in order to be effective.

The main indicator for management performance in private industry is profitability. Typically profits contributed by the facility will result in advancement and pay bonuses for the facility manager and thus he/she will endeavor to have a staff that will manage the various functions that contribute to an increase in profits and that avoid factors that may decrease them. On the positive side of the balance sheet are direct factors for increased profits: increased sales and lower production costs. On the negative or neutral side are indirect factors such as safety and environmental non-compliance, loss of production due to incidents and loss of reputation (although it could be argued that on the other hand a good reputation may lead to increased business and thus could be a positive factor). Maintenance is usually included in calculating the regular production costs, by balancing the cost of having the maintenance versus the cost of lost production due to equipment failures. Other factors that may not directly affect the facility's bottom line (at least not in the short term) but could have a severe effect on the facility manager's performance evaluation are safety and environmental impact on the community, personnel injuries and loss of life.

For the direct factors there are many leading indicators that the manager uses to forecast profit: net sales, orders, cost of raw materials and utilities, fixed costs (facility and labor), maintenance costs and non-compliance costs. We will use analogies to these indicators in the process safety area and propose indicators that can be used to forecast good process safety management.

EXISTING INDICATORS FOR PROCESS SAFETY

Lagging Indicators

Some facility managers still rely on lagging indicators such as an injury index based on hours worked at the facility, e.g. the OSHA 300 Index (29 CFR 1904), or the number of incidents that occurred in the facility in the previous period with an arbitrary target being set to indicate good safety performance. As it became evident that this indicator correlated well only with basic "hard hat" safety and almost not at all with significant incidents, the search for better indicators started. Better definitions as promoted by the Center for Chemical Process Safety (CCPS) of what constitutes a Process Safety Incident (PSI) and indices that include total plant working hours (Process Safety Incident Rate or PSR) and severity of the incident (Process Safety Severity Rate or PSSR) [1] will potentially bring about in the future better indicators, although those will still be lagging indicators.

The use of "near misses" as an indicator has been tried (and is still being tried) with mixed success. Defining exactly what constitutes a near miss is difficult and the implementation of the program is fraught with problems—employees were reluctant to report too many misses for fear of appearing unable to perform their job or for fear of being blamed for them in spite of extensive support from management. When the definition of a near miss was narrowed to include only releases that exceed a threshold indicating a potentially significant process safety impact, the dearth of consistent data defeated the purpose. This is not to say that these efforts are without value because the involvement and emphasis of management when implementing these programs has brought about an improvement in safety culture in the facilities where they have been tried. This is though, a side effect which is observed frequently: people will tend to perform better when their performance is watched closely and the results meet the expectations. CCPS' efforts to "harmonize" industry-wide the definition of a near miss [1] may result in more consistent data down the line to assess the effectiveness of this indicator. But, it does not resolve the influence of human behavior on the data itself, i.e. underreporting or misreporting in order to look better.

With the advent of the OSHA Process Safety Standard (29 CFR 1910.119) the target became compliance with the Standard as a measure of good process safety performance. Although the implementation of the Standard has brought with it a reduction in the number of significant incidents in the petrochemical industry [2], compliance with it alone has not proven to be a good predictor of avoiding disaster. It is widely accepted that because the Standard is open to interpretation, the manner in which the elements of the Standard are implemented and execution follow-up are crucial to good process safety. There is a need for metrics to measure how well each of the elements are being performed (positive factor) or how many missteps are occurring that indicate impending systems failure and a possible severe incident (leading indicators). Since there aren't yet objective standards by which good performance can be judged, we are relegated to measure the deficiencies, or using the Swiss Cheese Model, the holes in the cheese. These can provide some leading indicators though only by going through an arduous road.

Leading Indicators

CCPS has proposed to use and standardize on metrics in the following areas [1]:

1. Maintenance of Mechanical Integrity
2. Action Items Follow-up
3. Management of Change (MOC)
4. Process Safety Training and Competency
5. Safety Culture.

These were selected based on the assessment of hazards inherent in operations, critical factors that resulted in major incidents, and metrics in the CCPS risk-based safety book. Except for Safety Culture, they are very detailed, and seem to require the involvement of experienced process safety professionals and the availability of sufficient resources in terms of instrumentation and personnel to collect them. The purpose is first to try to standardize on the definition of the metrics across industry and then collect the data that will lead to an understanding of the factors that result in good process safety. In order to achieve the latter, results will have to be correlated with lagging metrics such as actual and near miss incidents.

A paper sponsored by the Abnormal Situation Management (ASM) consortium [3] provides a good discussion of these metrics and suggests addressing other causal categories for metrics that should be collected in order to complete the understanding of contributors to process safety incidents.

CCPS' metrics proposal is timely and necessary to arrive at a technical understanding of process safety. It, like financial and accounting principles, will ultimately become the technical basis for guidance to management. That guidance is still yet to be developed and we are proposing a method to do so in a way that will get management engaged in process safety.

MANAGEMENT GUIDANCE

Unlike in the financial field, management is currently left without specific mid-way guidance in the process safety area. There will be an ultimate target once the metrics are accepted by industry, the information gathered, and the results correlated with good performance over time. The immediate guidance remains continued compliance with the Process Safety Standard, improved execution of its elements, and implementation of new concepts such as Operational Discipline and Safety Culture that have been developed in the last few years and seem to be harbingers of better process safety. The interpretation of how to do this continues to be left to the process safety professionals in the plant and participation of management is relegated to giving support to his staff without direct involvement.

In addition, developing the metrics proposed by CCPS requires a level of trained personnel, equipment and instrumentation resources, that many small and mid size companies won't be able to mobilize. When companies already have problems finding experienced people to review the safety analysis made in an MOC, and when there aren't enough resources to close the MOC that was started months ago, identifying and analyzing the type of changes that tend to bypass the MOC procedure (one of the CCPS metrics) may not be the highest priority for the facility. The facility manager will find it difficult to allocate resources for metrics while being unable to satisfy the usual requirements of PSM. Thus, simple management indicators are proposed where the metrics are collected as part of the usual conduct of business.

PROPOSED MANAGEMENT INDICATORS

The proposed indicators are simple in order to achieve the objectives of (a) easy gathering of the information, (b) easy communication to both employees and top management and, (c) a reflection of the level of risk in the facility at any point in time. If these objectives can be met, then site management will be involved in the administration of site process safety and the employees will be engaged. As mentioned above, just the fact that management gets involved tends to produce better results, in this case probably an improved safety culture.

The indicators are based on maintaining low facility risk as will be explained later. The areas of focus are:

1. Action Items Follow-up
2. Management of Change (MOC)
3. Process Safety Training, and
4. Operational Integrity.

These four areas are not too dissimilar to those chosen by CCPS as far as selected areas but there are differences in their treatment. Mechanical Integrity (MI) is not included as part of the metrics because this is an area where site management is already involved and, as discussed before, the objective of keeping the plant running without equipment breakdowns has already been set. It can be argued that the usual MI program will not single out the critical equipment and therefore it will not get the attention it deserves. But, let's give some credit to the people managing the program in that they understand the importance of the usual safety devices (reliefs and vent devices, interlocks and safety shutdown systems, emergency response equipment, etc.) and provide the appropriate care and inspection. As for other critical equipment, it is better covered under Operational Integrity (see below). The metrics in the other areas and the reasons for their selection are as follows.

Action Items Follow-up

Since our objective is to maintain low overall risk, any open action item from a PHA, an open incident investigation recommendation or an incomplete incident investigation impede attaining that objective. The metric here is the total number of open action items in each of the three categories. These numbers should be easy to obtain and should be reported on a monthly basis. The selection of this metric is obvious given the reason for performing PHAs and doing incident investigations—a potential risk has been discovered from the PHA on one side, or an actual risk exists and resulted in an incident on the other side, and that incident needs to be investigated. If the investigation was conducted and the action items were not closed out, that means that the causes of the incident have not been removed and the potential for an equal or similar incident is very real. It is also assumed that the PHA was conducted using a risk matrix and the action items resulting from it have already been screened for high risk. The percentage of action items completed is not as important as the total number of open ones because the percentage only reflects the resources put to work in that area—it is a measure of the rate of reduction of risk rather than the total risk present.

Management of Change (MOC)

By definition MOC represents flux in the plant outside of predefined safe limits. The more undergoing changes there are at a given point in time, the higher the total facility risk. That is, the more open MOCs there are, the higher the probability of making mistakes, if nothing else because the process safety information is fluctuating and the knowhow has not yet kept up with it. Furthermore, we know that startup or shutdown of a process represent a more hazardous operation than steady state. By analogy the sum of changing conditions in the plant increase the hazards of the plant. Therefore we want to know the total number of open MOCs at a point in time. If the number is high (facility dependent) it means either that the facility is engaging in too much activity, or if the changes are necessary, not enough resources are being allocated to complete the changes. This is the type of decision that site management ought to be making and where the underlying reasons are not difficult to fathom. The number will depend on the size and dynamics of the facility. It should be reported monthly.

The corollary to risk represented by change is that the longer a change is incomplete the higher the probability of something going wrong. Since we already require an expected date of completion when doing an MOC, we need to watch that that date is not exceeded. When dealing with temporary changes, this becomes more critical because we have done an assessment on the safety of the change taking into consideration a specific and limited length of time. Thus, it follows that both the number of past-due MOCs and past-due temporary MOCs ought to be monthly metrics for risk reduction. Collection of these data ought to be simple since they should be part of any efficient MOC system [4].

Process Safety Training

Having well trained personnel to operate the plant processes is essential to good process safety. It is assumed that a plant will have competent trainers and a method for testing effectiveness of the training. As opposed to CCPS' metric in this area, the measure that is being sought is one of training of anybody that operates a process rather than just those in PSM critical positions. In order to maintain a continuous level of skills, refresher training is required and is mandated by the PSM Standard. Because of the nature of training and the capacity of an individual to absorb the training, it cannot be a once-a-year event but needs to be distributed throughout the year. To follow if this is being done, the metric would be the percent of operators that have been provided refresher training in a quarter on one-fourth of the existing operating procedures. The objective would be to provide refresher training to all the operators and maintenance personnel on all their procedures at the end of each year. By watching this number, the site manager can track the maintenance of skills of the work force.

Operational Integrity

The metrics in this area try to capture the quality of operations that have inherently higher risks. This is accomplished by inspection of the PHA completed for the process and identification of the personnel, equipment and instrumentation that act as safeguards against the occurrence of a high-severity incident. This personnel, equipment or instrumentation would then be considered "critical" and a maximum time for their unavailability and/or backup protections for the missing item would be defined. For example, a redundant flammability meter used to initiate a shutdown (1oo2) that is out of order could be out for a day until a replacement is found and installed before having to shut down the process. A backup during the unavailable time could be setting the flammability alarm at a lower setting and/or putting an additional operator to watch it (if the same instrument was out for an hour for routine calibration this should have been taken into consideration when assessing the risk during the PHA and wouldn't constitute a critical unavailability).

In the same vein, if there was a critical administrative safeguard, e.g. operator rounds, in which a variable is read at a certain frequency during a shift, then a decrease in coverage during that shift due to an operator's absence would be a critical unavailability. The metric in all cases would be the number of "critical unavailabilities" recorded each week. The unavailability would be recorded even if a backup plan was in effect. Of course, just recording the unavailability wouldn't excuse not taking corrective action before too much time passes. The number of unavailabilities during a period would be an indication to management of how well the Mechanical Integrity program takes risk into consideration, of how good the process risks are understood by the operators, and operations supervision understanding and follow up in maintaining operational discipline to minimize the length of the outages.

Once the upfront definition of the critical items has been completed, collection of the metric should be straightforward. Unavailable critical items would be posted on a high-visibility board in the control room and remain there until those items are returned to service. This would serve the dual purpose of reminding operators of the unavailability of these items during and across shifts, and help record the unavailabilities for each shift. The daily numbers would be reported to site management possibly on a bi-weekly basis depending on the facility.

CONCLUSIONS

Simple indicators of facility risk have been proposed for use by site management for making decisions that impact process safety and for communicating the health of the process safety program to employees and top management. These indicators are:

1. Action Items Follow Up – the number of open action items from PHAs and incident investigations and the number of incomplete incident investigations;
2. Management of Change (MOC) – the number of monthly open, past due and temporary past due MOCs;
3. Process Safety Training – the percentage of operations and maintenance personnel given refresher training in each quarter to ensure that all personnel will be trained on all procedures within a year's time; and,
4. Operational Integrity – the number of daily critical unavailabilities.

These indicators do not claim to be a measure of total facility risk but they provide direction on how to reduce and maintain a lower facility risk and are predicated on a good underlying process safety program. Because of their easily understood relationship to risk, the indicators will make site management more apt to become involved and directly influence the performance of process safety in the plant. The indicators provide a continuous short-term link between desirable management action and the underlying technical basis without having to have total reliance on the technical staff.

Direct involvement of management in process safety can only improve Safety Culture in the site which is turning out to be a key element in good process safety. Additional improvements can be obtained by integrating these simple metrics into the management structure, i.e. a management system designed with process safety in mind [5]. It is hoped that with the continued use of these indicators as a communication tool will help to build in process safety performance as a part of total site management's performance and eventually be rewarded similarly to fiscal performance.

REFERENCES

1. Center for Chemical Process Safety (CCPS), "Process Safety Leading and Lagging Metrics – You Don't Improve What You Don't Measure", CCPS Website Publication, 2008.

2. Russ Elveston, former OSHA Area VI Assistant Director in his presentation at the STS-AIChE PSM Workshop, Houston, February 6, 2009.

3. Peter Bullemer and Doug Metzger, "CCPS Process Safety Metrics Review: Considerations from an ASM (Abnormal Situation Management) Perspective", 11th Annual Symposium Proceedings, Mary Kay O'Connor Process Safety Center, p.423, October 2008.

4. Jack Chosnek, "An Effective Management of Change (MOC) System", STS-AIChE Workshop, Houston, November 2, 2006.

5. Jack Chosnek and Ricky Clifton, "Improved Process Safety Management and Simple Metrics", Process Safety Progress, Vol. 27 No. 4, p. 284, December 2008.

What Does Management Need to Know To Lead Process Safety Improvement?

Mark Paradies, System Improvements, Inc.
238 S. Peters Road, Ste 301, Knoxville, TN 37923
http://www.taproot.com
865-539-2139

ABSTRACT

There are five keys that managers need to understand to improve process safety:

1. Management's effect on performance improvement.
2. How performance improvement and root cause analysis work together.
3. How to change behavior effectively.
4. How to move from being reactive to being proactive.
5. How to use effective trending to assess data and manage process safety.

This paper explains a little about each key and provides examples of how the failure to understand a key has lead to poor performance. It will address the knowledge needed by Division level leadership and Plant Managers of large industrial facilities.

INTRODUCTION

Process safety management has been around since the early 1990s. It was created in response to several major chemical plant accidents. The goal was to stop the tragic, often fatal, and environmentally damaging releases of hazardous substances. Despite almost 20 years of efforts, accidents continue. This leads people to ask:

"*Why isn't our company able to sustain improvement?*"

The answer is usually simple. Senior management doesn't know how to lead process improvement – which includes leading improvement of process safety.

In this paper, we will address the knowledge needed by Division level leadership and Plant Managers of large industrial facilities.

FIVE UNKNOWN KEYS

In talking to senior management and observing performance improvement programs, the five key things this level of management should know (but often doesn't) are:

1. Management's effect on performance improvement.
2. How performance improvement and root cause analysis work together.
3. How to change behavior effectively.
4. How to move from being reactive to being proactive.
5. How to use effective trending to assess data and manage system safety.

Some might say that these keys are obvious. Any manager worth his pay should know these keys and apply them to improve performance – including efforts to improve process safety. Yet, the examples that follow show that these keys are often missing and can lead to expensive, even fatal, consequences when ignored.

Key 1: Management's Effect on Performance Improvement

What is management's effect on a performance improvement program? If you ask the people in the field, you will hear the answer. Many managers talk a good game. They pledge their commitment to safety, process safety, quality, Lean, and environmental sustainability. They buy posters, issue memos, and talk a good talk. But people in the field know if they walk the talk.

For example, people in the field know the answers to these questions:

1. Do managers know what they are talking about? Do they understand the details of the program and what it takes to use the improvement tool? Did they take the training they ask others to attend? Or are they aloof from the program?
2. What do managers ask for? Do they talk about performance improvement but demand that production numbers are met above all else? What do they ask about *FIRST*? What is rewarded in management's bonus program?
3. Where does the budget go? Are performance improvement funds cut first in difficult times?
4. Where do managers spend their time? Where do they focus? What data is tracked and monitored? Do they actively participate in process safety activities?
5. Who is assigned to what jobs? Is leading the performance improvement/process safety effort seen as a plumb assignment that leads to promotion?

6. Does senior management insist that their subordinates are trained in the performance improvement/process safety techniques? Or is process safety seen only as a job for safety?

How hard is it hard to judge management's impact on process safety? Let's look at a real example of the biggest refinery in the United States. Prior to the BP Texas City Refinery ISOM explosion[1]:

- After several years of pre-merger budget restrictions (cost cutting getting ready to be sold), senior management, lead by the CEO, ordered a post-merger 25% across the board budget cut.
- In 2002, a flare tie-in for the ISOM was cut to save $150,000 on a capital project.
- The backlog of corrective actions for hazops and incident corrective actions had items in it over a decade old. Management was not tracking or acting to reduce the backlog.
- "Good people" were assigned to key safety jobs but they did not have adequate technical training (for example, no PSM training for the HSE Manager).
- BP Texas City had fatalities in the year prior to the explosion, but the Plant Manager and his key subordinates thought they were improving safety because lost-time stats "looked good" compared to industry averages.
- Management approved buying "cheap trailers" (not blast resistant) to save money. Later, the permit to site the non-blast resistant trailers next to a process unit (the ISOM) was approved.

The US Chemical Safety Board[2] and/or attorneys representing blast victims[1] saw these management actions directly linked to the accident that followed. I assume that the senior managers did not understand the effect on process safety. After all, they all (all the way up to the CEO) testified in trial or in depositions that they didn't think these decisions were directly linked to actual performance (the explosion)[1].

Key 2: How Performance Improvement and Root Cause Analysis Work Together

Senior management needs to be the sponsor of any performance improvement program, especially one to improve process safety. As the sponsor of the program, they need to understand how the program works, including how root cause analysis is a key part of a performance improvement. To understand how to use root cause analysis in a performance improvement program (both reactively and proactively), senior management must understand root cause analysis.

Again, let's look at a real example - the management of the BP Texas City Refinery prior to and immediately after the ISOM explosion:

- The Plant Manager sponsored a safety culture survey (the Telos Report) but there was no ownership from above (more senior management) for corrective actions to the problems detailed in the report.
- Senior management didn't know what was wrong with the root cause analysis techniques being used at BP. Shortcomings were evident in the BP Texas City explosion Interim Report[3] and the Final Report[4].
- Senior management was stuck in and promoted a blame culture (counter to good root cause analysis). This was evident in the statement/press release after the BP Texas City explosion. In that statement, Ross Pillari, President of BP Products North America, said[5]:

  > "Had the six operators and one supervisor assigned to the start-up of the refinery's so-called isomerization unit been doing their jobs, the explosion would not have happened, 15 people would not have been killed and more than 170 would not have been injured."
  >
  > "The mistakes made during the start-up of this unit were surprising and deeply disturbing."
  >
  > "The core issue here is people not following procedures."

- What action was taken by the management team before the Interim Report was even released? Several operators and supervisors were fired. However, even after a critical internal report (the Bonse Report[6]), no senior managers were fired (although the plant manager was relieved of his job and put on paid leave and all senior managers above him through the CEO – for various reasons - are no longer at BP).

Ask yourself: Are the actions of BP senior management unusual? Or would this type of action be seen as typical? I believe if they understood advanced root cause analysis, they would not have taken these counter-productive actions.

Key 3: How to Change Behavior Effectively

One of the key challenges that management faces is changing behavior. In all poor performing plants, including those with poor process safety, bad behaviors of employees will be an issue. How management changes behavior will have a major impact on the success or failure of the improvement initiative.

Management will have to effectively use discipline to change behavior. But, even more important, management must also understand how to use soon-certain-positive rewards – positive discipline – as a behavior change tool.

Management may forget that employees constantly look for ways to save time and effort. Also, employees respond to the actions of supervisors to enforce the rules (or turn a blind-eye to the rules when they conflict with production priorities). Finally, although official bonus systems are noticed, unofficial bonus systems (peer recognition, supervisor attaboys, risk taking thrill, relief of boredom, and looking good) may be even more important.

Your Example	Old Behavior:	New Behavior:
Time		
Effort		
Enforcement		
Bonus		

Figure 1: Behavior Change Matrix
(Copyright 2008 by System Improvements, Inc[7]. Used by permission.)

In our root cause analysis training, we teach people to understand how to change behavior using a Behavior Change Matrix (Figure 1) to develop corrective actions. The Matrix helps investigators (or managers) understand the factors that caused the old, undesirable behavior and develop soon-certain-positive methods to reinforce a new, desirable behavior.

Using negative discipline policies are at least as old as the days of sail in marine transportation. In the 1980's W. Edwards Deming declared them obsolete when he made "Drive out fear ...[8]" one of his "14 Points" for management to transform a company. But to this day, a majority of companies use ineffective late-uncertain-negative discipline programs to change behavior. Therefore, I must conclude that *management must be unaware* of how to effectively change behavior.

Key 4: How to Move from Being Reactive to Proactive

A majority of companies are reactive. They respond to accidents, incidents, and problems rather than proactively monitoring and improving performance. Proactive improvement prevents accidents by improving performance before the accident occurs. This avoids the many negative consequences of accidents and is highly efficient by avoiding waste – a concept that is an integral part of the Lean management philosophy.

What happens if management remains focused on reactive performance improvement? The nuclear industry is a typical example.

In the nuclear industry, the bigger incidents are investigated well. But the smaller, less significant incidents (of which there are thousands) are investigated poorly using Apparent Cause Analysis. Despite the poor analysis, corrective actions are usually developed for the less significant issues. Thousands of corrective actions for the less significant issues result in a backlog of corrective actions and wasted effort implementing fixes for problems that really don't deserve to be fixed. Also, the backlog includes fixes for more significant problems that may not get attention because of the size of the backlog.

In addition, the data from the poorly analyzed small incidents is included in the database with the significant incidents. Because of the sheer numbers of the smaller incidents, trends in significant issues are lost in the noise of the data.

Nuclear management - without the ability to detect significant trends and reacting to old, insignificant, backward looking data – is surprised when bad things happen. They are whipsawed from thinking performance is great to reacting to negative regulatory findings or poor plant performance.

If this sounds familiar, that's because a majority of companies are run in the reactive mode. Therefore, I would assume that managers don't understand how to collect data and manage proactively – especially in their process safety management programs.

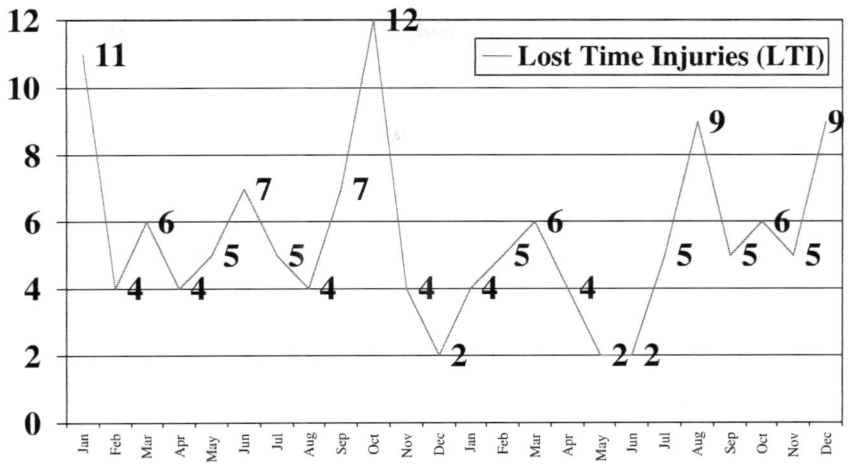

Figure 2: What Months Are Exceptionally Bad?
(Copyright 2008 by System Improvements, Inc[7]. Used by permission.)

Key 5: Using Trending to Assess and Manage Process Safety

Management pays attention to trends and uses them to decide what to do. Unfortunately, many managers have knee-jerk reactions to trends they see as "adverse" when the data was actually the result of everyday, normal variation. For example, see Figure 2.

Think of time progressing. When October had 12 injuries, the senior manager reacted. He fired the safety manager. He hired a safety consultant. He told all this managers to hold a safety stand-down. And he put out a memo that said that safety was a condition of employment (watch out – you too could be fired).

And the next month he was pleasantly rewarded. Only four injuries. The month after that? Even better. Just 2 injuries. Surely, this proves that his reaction was correct.

In reality, the 12 injuries were within what should normally be expected from the process. If a Process Behavior Chart was drawn (as in Figure 3), you could see that the process normally produces somewhere between 0 and 12.5 injuries per month (average of 5.5). And what about the decrease in injuries after his overreaction? It can probably be attributed to normal regression to the mean and under-reporting due to fear of being fired. It certainly didn't produce long tem improvement (the process continued on well within the expected bounds of zero to 12.5 injuries per month).

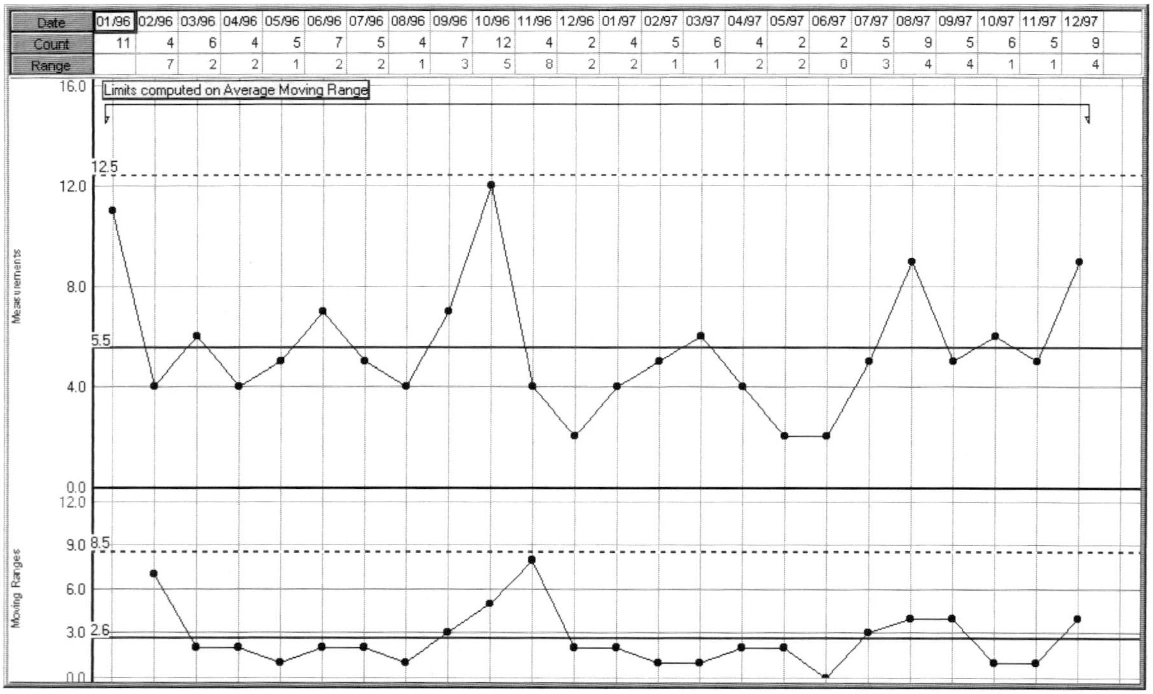

Figure 3: Same Data on a Process Behavior Chart
Copyright 2008 by System Improvements, Inc[7]. Used by permission.

Walter Shewhart[9] proved that over-reacting to perceived trends actually made processes behave more unpredictably (worse) than if no action was taken at all. Let's look at this fact in light of management's action in the example above. Could the over-reaction actually make things worse?

First, firing the safety manager is not going to make it easier to find a well-qualified replacement. This could result in an unqualified, but hard-nosed replacement who thinks that firing people who make mistakes is the best way to improve performance. When this is combined with the memo that says "Safety is a condition of employment...", people will probably be less willing to report safety problems and more likely to cover up smaller incidents (to avoid discipline).

When incidents and near-misses aren't reported, problems can't be corrected until something major happens. Thus, big accidents aren't prevented until they happen. This could lead to an unexpected future major accident when management thought everything was going along fine. And that is how a process becomes more variable after an over-reaction to a "non-trend".

Despite Shewhart's and Deming's work and the popularity of Six Sigma, many companies still over-react to perceived trends. Thus, using trends to assess and manage is still something that management needs to learn to successfully manage process safety improvement.

CONCLUSION

Even though process safety management has been around since the early 1990's, we are still having unacceptable accidents. We are so close. If only management could recognize the keys needed to achieve improved process safety, they would achieve the performance desired. If not, they will continue to experience uneven, erratic safety performance despite spending time and money trying to achieve process safety.

REFERENCES

1. "BP Texas City Plant Explosion Trial Declassified Evidence Release," http://www.texascityexplosion.com, verified 1/30/2009.

2. U.S. Chemical Safety Board, "INVESTIGATION REPORT, BP, Texas City, Texas, Refinery Explosion and Fire," March 2007.

3. John Mogford, "Fatal Accident Investigation Report (Interim Report)," BP, Texas City, TX, May 12, 2005.

4. John Mogford, "Fatal Accident Investigation Report," BP, Texas City, TX, December 9, 2005.

5. "BP Blames Staff for Blast," *Houston Chronicle*, May 18, 2005.

6. Wilhelm Bonse-Geuking, et. al., "Management Accountability Project, Texas City Isomerization Explosion Final Report," BP, February 2007.

7. Mark Paradies and Linda Unger, *TapRooT® - Changing the Way the World Solves Problems*, System Improvements, Inc., Knoxville, TN, 2008.

8. W. Edwards Deming, *Out of Crisis*, Cambridge University Press, Cambridge, MA, 1988.

9. Walter Shewhart, *Economic Control of Quality of Manufactured Product*, D. Van Nostrand Company, Inc., USA, 1931.

63e Process Safety – Perspectives from the Top

Thomas Archibald
Rohm and Haas Company
Philadelphia, PA 19106
thomasarchibald@rohmhaas.com

Abstract:

Mr. Tom Archibald, Vice-President and Director of Operations at Rohm and Haas Company will highlight what information executives need, expect, and prefer from process safety professionals and from managers for decision making and resource allocation to address process safety and process risk issues. He will also provide guidance on how that information can best be conveyed to upper management and executives within an organization.

Topics addressed will be based upon his experiences and may include: what keeps executives "awake at night", what is needed to make critical decisions, what are key components of the information needed, what level of detail may be required and how it can best be presented, what is included in a winning "pitch" to executives, use of financial information, and dealing with process safety and risk in a cyclical economy.

Case studies may be used to demonstrate process safety or risk decision examples - "positive" ones where the analysis and decision process went well, and "negative" ones where things went awry - and what can be learned in each case.

Procter & Gamble Global Risk Ranking

Frank H. (Hank) Gurry, The Procter & Gamble Company
8256 Union Centre Boulevard West Chester, Ohio 45069
gurry.fh@pg.com

and

Robert W. Johnson, Unwin Company
1920 Northwest Boulevard, Suite 201, Columbus, Ohio 43212
rjohnson@unwin-co.com

ABSTRACT

The Procter & Gamble Company is on its third generation of company-wide relative risk ranking models, the first of which was implemented over 20 years ago. The data from these models have been used to improve systems and reduce risks at sites throughout the Company. Both of the earlier models focused on consequences of incidents, and were heavily weighted toward off-site consequences of chemical releases. Over 95% of the high-risk systems in the 1988 model from this type of incident, such as systems using oleum or chlorine, have been eliminated or replaced with inherently safer systems.

More recent incidents involving large property losses but with no direct community impact motivated a scope expansion of the risk ranking effort. The current model now includes nearly 600 systems at 108 Procter & Gamble facilities worldwide. In addition to systems with the potential for toxic and/or flammable releases, the model includes special known risks such as drying operations, frying, perfume making, and systems with dust explosion or bursting vessel explosion hazards above thresholds of concern. This presentation shares features of the current risk ranking model, describes challenges faced during model development and data collection activities, and shows how business units have made decisions that drive down and eliminate risk. Selected sections of the survey form used to collect system data are included in the appendix to the paper as examples.

INTRODUCTION

Procter & Gamble (P&G) is a multinational consumer products company, with over $80 billion in annual sales, and with about 140 manufacturing plants in over 40 countries functioning across four geographic regions. The Company is organized into three primary business units – Family Care, Household Care, and Health and Beauty Care. Within this structure, the company has 44 brands with sales of $500 million or more. The scale and scope of the business and brands, the geographic reach, and the broad range of technologies in manufacturing operations make reliability a key factor in the product supply function. It is within the context of employee safety, community impact, and reliability that P&G has developed and used relative risk ranking models as a tool in making decisions about risk management.

1988 RISK RANKING MODEL

P&G has an excellent history of safe operations. One of the tools we have used to support decision making and improvement in manufacturing operations is the development and use of relative risk ranking models. The first model was developed with Battelle in 1988, and it modeled the potential risk associated with chemical exposure to people off-site from an accidental one-time release. The 1988 risk ranking model measured the relative risk to the neighboring communities from 68 toxic and fuming chemical systems, such as sulfur trioxide, sulfur dioxide, bromine, chlorine, and oleum. The model included all geographic regions. It used data such as quantity of material unloaded and stored, frequency of unloading, line sizes, system pressure, and community population location and density.

Ten years later, in 1998, eight of the top ten systems with the highest relative risk ranking had been eliminated and the overall fuming chemical risk to off-site populations had been reduced by 95%. While this risk reduction was achieved in part through a combination of business decisions such as plant consolidations, divestitures, and formula changes, the risk ranking model and potential safety and community exposure data also played a part in system improvements that led to reduced exposure levels.

1998 RISK RANKING MODEL

In 1998, also working with Battelle, P&G developed and used a second-generation relative risk ranking model. In addition to fuming and toxic chemical systems, this model included estimates of relative risk from bulk flammable liquid systems, other highly hazardous chemical systems, large pressure vessels, combustible dust handling systems, and a number of special known risks such as spray towers. The 1998 model ranked 290 systems in all regions, and the risk ranking included both off-site and on-site risks and exposures. The model was based on input from a detailed questionnaire completed by each plant, and reviewed by regional and corporate resources. It provided an order-of-magnitude comparison of risk for a broad range of operations.

Within the product supply organization, the model results were used as a guide for businesses in their efforts to continuously improve operations, allocate scarce resources, and reduce risk. By 2005, the model had helped drive a reduction of 70% of the total relative risk of the systems included in the model, and it helped drive a reduction of 50% of the risk associated with chemical systems. As with the earlier model, other business decisions and other risk management tools such as internal health, safety and environmental (HSE) audits, third party audits, risk assessments, etc., played key roles in risk reduction.

CURRENT RISK RANKING MODEL (2007 MODEL)

Data from the first two risk ranking models were used to improve systems and reduce risks at manufacturing plants, primarily by focusing on off-site consequences of chemical incidents. During the period of 2003 to 2006, the company experienced a few large incidents, some of which are shown in the following pictures. These incidents primarily involved on-site property damage, and some had the potential for business interruption. They did not result in off-site exposures, and had no recordable injuries. Incident investigations revealed that procedures and training for start-up, shutdown, operation, maintenance, and equipment cleaning contributed to some of these losses.

Figure 1. Failure in a Chemical Processing System

Figure 2. Fire in a Drying System

Figure 3. Failure of a Co-generation System

Figure 4. Oxidizer Decomposition

In early 2006, P&G management decided to evaluate current systems to better understand risks and set priorities for improvement. Working together with Unwin Company, we developed the 2007 risk ranking model. The objectives of the model were as follows:

1. Identify and rank systems in relative order of risk.
2. Raise awareness and understanding of these risks.
3. Take appropriate action to reduce risks.

In the 2007 model, as in the two previous models, *risk* is defined as the product of severity of consequences ($/incident) and likelihood of occurrence (incidents/year). For each system, the model calculates a risk priority number (RPN), and then ranks all the system RPNs in relative order. Business units can then compare the relative risk of systems within their business unit, or can compare their systems against systems across the entire Company, and use this data to help make decisions.

While off-site exposure is still very important to manage well, the 2007 model is designed to give additional focus to on-site property damage from large incidents. The model is not part of internal company HSE audits of plants, and it is not a factor in determining internal HSE audit scoring and actions. Rather, it is intended to be a separate tool to help business units make risk management decisions across the many plants and systems within the business unit.

Data Collection

The model starts with a detailed questionnaire or survey form that is completed by people at the site. Some sections of the survey are included as an Appendix to this paper. The survey form was available in both metric (SI) and US units.

Data is collected both for the site and for the specific system in the following general categories:

- Typical on-site and off-site population numbers, density and distances
- Environmental data such as temperature, wind, elevation, topography, and distances to waterways
- Chemical information, including chemical properties, inventories, unloading frequencies, pressures, line sizes, containment features, detection systems, and control systems
- Estimates of potential property damage and business interruption from an incident involving that system
- An estimate of the frequency of a large incident on a particular class of system, using both historical Company data and some industry data

- An evaluation of the effectiveness of plant systems such as management of change; incident investigation; process safety studies; process hazard analyses (PHAs); testing, inspection and maintenance systems; operation and maintenance procedures; training and qualification systems; and emergency action and response systems.

Major classes of systems evaluated in the model include:

- Chemical systems (unloading/storage systems; transfer/processing systems)
- Potential dust explosion systems
- Potential bursting vessel hazards
- A group of special known risks identified by business unit and corporate resources based on Company experience

In some cases, decision flow charts are used to help guide the plant and business unit in deciding whether to include a system in the model, and how to answer questions in the questionnaire. Resources from the business unit and corporate resources review the data and work with the plants in an iterative process to align data on similar systems. The Excel®-based model then takes data from the input questionnaires and provides a relative ranking of the risks of business interruption, property damage, environmental impact, and on-site and off-site injury for the four major classes of systems.

One strength of the overall HSE program at P&G is the use of standard methods throughout the Company. These include standard methods for elements such as completing process safety studies, HSE audits, training and qualification of site HSE program leaders, treatment of highly hazardous chemical systems, incident investigation and reporting, etc. Use of common HSE tools and procedures, metrics, audit protocols, and data systems facilitated the data collection process for sites and helped with subsequent comparison of systems within and between business units.

Model Development

The following is a timeline for the 2007 model:

- February 2006 – Project started with agreement on overall goals and strategy
- February to April 2006 – Excel-based model developed by Unwin Company
- April to June 2006 – Model pilot-tested on ten systems from various business units and regions
- June to September 2006 – Plants collected data on about 450 systems and data were entered into the model

- September 2006 to January 2007 – Business units, plants and corporate resources reviewed data and results, checked for alignment and made changes as needed
- April 2006 to January 2007 – Working with Unwin Company, results were reviewed and model refinements were made
- February 2007 – Data presented to management; areas and systems identified on which to focus risk reduction efforts
- 2008 – Model expanded to its current size of almost 600 systems.

Emphasis of the Current Model

Several of the incidents that led to the development of a new risk ranking model involved property damage (PD) and potential business interruption (BI), but did not involve many highly hazardous chemical systems or off-site exposures. In addition, a review of incidents showed that gaps in procedures, training, operation and maintenance contributed to these incidents. The analysis of incidents suggested that a new risk ranking model should include systems with potential for PD and BI, in addition to the review of chemical systems and off-site exposure that had been the focus of previous models.

The 2007 Risk Ranking Model also added some new factors to give additional emphasis and focus to plant operating and maintenance procedures and training and qualification. These factors, called likelihood modifiers, are broken into two site-specific and eight system-specific modifiers. Using a table of descriptors (as shown in the Appendix, such as Section 2.5 of the Survey Form), together with data from annual HSE audits, plant and business unit resources evaluated the current status of each system. Each likelihood modifier is assigned a weighting in the model, and the combined weighting of the likelihood modifiers has a strong influence on the final risk priority number (RPN) for that system. Thus, the current status of elements such as management of change, operating and maintenance procedures, training and qualification, and testing and inspection, etc., are important factors in determining the relative risk ranking of a given system.

Results to Date

Using the risk ranking results from the February 2007 model, business units and plants developed plans and actions to reduce the risk from higher ranked systems. The one-year results from these risk reduction efforts, as predicted by the model, are summarized below:

- Total Company risk reduction of 26%
- Reduction of the frequency of large incidents of 25%

- Significant improvement in plant operation and maintenance systems
- Focus and progress on process systems having a combustor associated with the process
- PHAs completed on a number of systems which previously did not require a PHA.

Additional Details of the 2007 Model

A tremendous amount of effort has been invested in model development, data collection and ensuing risk management efforts. This has obviously involved a long-term commitment on the part of P&G's corporate management and HSE resources. Some of the details of model development and content are given here for insight into the level of detail associated with the 2007 model.

Excel-Based Model

The earlier 1998 model was a compiled program with an Access®-based front end for data collection and validation. The decision to convert to an Excel-based model was made primarily for reasons of model transparency, since the earlier model was viewed as a "black box" that model users were unable to "peer into" for an understanding of how the different parts of the model worked and what factors go into the risk calculations. The Excel-based platform has also facilitated refinement of the model, validation of the model data, dissemination of the model results and communication among model users.

A detailed program description of the earlier 1998 model made it possible to use it as a starting point for 2007 model development. The additional features such as the likelihood modifiers were then incorporated as model development progressed.

The Excel-based model makes it possible for nearly all of the calculations to be "live"; i.e., whenever a data input is changed, the effect on all of the results are seen immediately. The one major part of the earlier model that was not incorporated into the Excel-based model was the plume calculations. A decision was made to use the "ALOHA" model (Areal Locations of Hazardous Atmospheres), disseminated by the U.S. National Oceanic and Atmospheric Administration (NOAA) and U.S. Environmental Protection Agency (EPA), for both toxic and flammable plume calculations. Thus, any change in a parameter affecting the ALOHA inputs necessitating re-running the ALOHA model and updating the corresponding results in the risk model.

The final Excel-based model, with all data incorporated, is about six megabytes in size and includes fourteen spreadsheet tabs:

- Home – model version; risk summary by global business unit; global parameters and assumptions
- Results – all intermediate and final risk results calculations, including source term calculations (111 fields calculated for each of the ~ 600 systems)
- SiteData – site-specific input data (21 data fields for each of the 108 P&G manufacturing facilities having one or more systems in the model)
- SysData – system-specific input data (87 data fields for each of the ~ 600 systems, not all of which are applicable to each type of system)
- PlumeData – all input data used by the ALOHA model that is not already in the ALOHA model database, plus all output results from the ALOHA model runs (48 fields for each of the 200+ systems requiring plume calculations)
- ChemData – chemical-specific input data (22 data fields for the 36 chemicals in the model having flammability, inhalation toxicity and/or ecological toxicity properties)
- SKData – data for the 19 different "special known risks" included in the model
- FreqData – base major-incident frequencies by hazard type and operation type
- Modifiers – likelihood modifier weightings
- StreamData – lookup table for stream flow rate as a function of channel type, slope, roughness, width and depth
- Results Key, SiteData Key, SysData Key, PlumeData Key – description of each input and output parameter, including units of measure and expected values.

Data Validation

As may be anticipated, entry and validation of all of the input data for the nearly 600 systems included in the risk ranking posed significant challenges that were managed by several means:

- As much as possible, data were exported from the 1998 model and imported directly into the new model after alignment of pertinent data fields
- Data from the earlier model were also used to populate survey forms for dissemination to the various business units, for updating of the existing data and collection of new parameters used by the 2007 model
- Any data falling outside of expected ranges or values were flagged for review
- Data for new systems not in the previous model were collected using the survey forms and their associated instructions
- These data were reviewed by the business unit HSE representative before delivery and by the model developer as the data were entered into the model

- Features of the Excel-based model made additional data validation possible by sorting and filtering in various ways, so that comparisons could be made such as between similar systems at different locations
- Both intermediate and final results were checked for reasonableness and for comparison between similar systems
- Updates were made throughout the process as the business units and facilities changed and improved systems and a few systems were decommissioned or otherwise eliminated from the model.

Frequency Estimates

The likelihood of a major incident for each system was estimated, for purposes of the risk ranking model, as an annualized frequency with four types of inputs:

- Base major-incident frequency for the specific hazard type and operation type, estimated by accumulated company experience and, where necessary, broader industry experience
- Product of all likelihood modifiers based on each of ten risk management systems being assessed as less than basic, basic, or more than basic
- Frequency modifier for systems involving unloading of a hazardous material, based on the frequency at which the unloading operation is conducted
- For a few systems having the potential for a boiling-liquid expanding-vapor explosion (BLEVE), a frequency reduction for systems having a well-maintained automatic deluge system and drainage with remote impoundment.

Severity Estimates

The severity of consequences of a major incident for each system is estimated, for purposes of the risk ranking model, as a combined impact on workers, off-site populations, business interruption, property damage, and the environment. For most impacts, the key parameter to be calculated is an exposure radius. The exposure radius is the distance to which effects could be anticipated at an ERPG-2 (Emergency Response Planning Guideline, level 2) concentration for inhalation toxicity or a comparable level for fire thermal radiation or explosion blast effects. For systems having inhalation toxicity hazards, an effect area was also estimated, since the effects are directional. Source term calculations include release rate and duration estimates, flashing, aerosol production and pool evaporation. System features could allow for vapor release mitigation taking into account the time required to detect and implement the mitigation systems and the anticipated effectiveness of the systems.

APPENDIX – Selected Sections of the SURVEY FORMS

Risk Ranking Model 2006 Input -- (Metric Units)

Fire, Explosion, and Chemical Release Risks are being ranked for all P&G sites globally for systems which can lead to Class IV technical safety incidents. Your help is greatly appreciated in completing this survey.

Provide your best estimate for each value. **Do not leave blanks**. If a question does not apply, please indicate "Not applicable."

SECTION 1

INTRODUCTORY GUIDANCE

1.1 INTRODUCTION

The Risk Ranking Model allows the relative risk of a particular system to be determined and then compared with other systems in the company. It looks at:

- acute chemical exposure to people on site
- acute chemical exposure to people off site
- loss of property from fire and explosion
- direct discharge of chemicals into open waters
- business interruption risk (lost profit) associated with fire, explosion and large mechanical failure.

The risk ranking model is a tool for identifying risk reduction opportunities, prioritizing risk reduction efforts, and evaluating new processes and acquisitions. The survey form is used to collect input data on a specific system. The model uses this input data to calculate a relative risk score for potential fire, explosion, and hazardous material release risks. These scores are then compared to develop risk ranking on a GBU, regional, and global basis.

The model will estimate the impact of the following incidents:

- a release of a chemical,
- a Boiling Liquid Expanding Vapor Explosion (BLEVE),
- a large fire in the process,
- an explosion in a pressure vessel or combustible dust system, and
- other special known P&G risks.

1.2 RESOURCES TO COMPLETE THE SURVEY FORM

This survey form should be completed by the HS&E leader for the site. Input from Environmental, Process Safety, IH&S, and Operations personnel will be essential to complete this form accurately. Tools such as: process flow diagrams, plant layout diagram, process safety studies and business continuity plans will also be useful. If you have any questions please contact your Technical Safety GBU Resource.

1.3 SURVEY FORMS

The Risk Ranking Tool is made up of 6 sections and survey forms:

- Site Data. Section 2.
- Chemical Hazard. Section 4.
- Dust Explosion Hazard. Section 5.
- Bursting Vessel Explosion Hazard. Section 6.
- Special Known Risks. Section 7.
- Likelihood Modifiers. Section 8

Guidance:
As you will see in the electronic survey forms the Likelihood Modifiers from Section 8 are integrated into the survey forms for the various systems. This is to ensure that your answers to the Likelihood Modifiers questions are linked directly to the system being considered. If you are completing the Microsoft Word version of the survey form first, before entering data into the electronic version, you will have to make copies of Section 8 for each system you are considering.

If a Chemical from the pre-identified list is stored in a pressure vessel, treat that as a Chemical Hazard and fill out the Chemical Hazard survey form (Section 4).

SECTION 5

DUST EXPLOSION HAZARDS SURVEY FORM

First use the flowchart in this section to determine if you have dust explosion hazards. Then complete a **Dust Explosion Hazard survey form** (starting in Section 5.1) for each room, building, or dust handling system that meets the criteria in the flowchart. Note: There are a number of Special Known Risks (namely Spray tower, Fluid bed dryer, Continuous coffee roasting, Coffee pre-dryer, PH&N Receiving/Milling and PH&N dryer) which are a dust explosion hazard but are addressed in Section 7. Therefore Section 5 should not be completed for these systems.

Guidance: A survey form should be completed for each system identified by the flowchart. For example if by following the flow chart a single dust explosion hazard is identified, e.g. accumulation in a building, fill out a single dust explosion hazard survey form. If by following the flowchart another hazard is found, e.g. an indoor baghouse filter, complete a second survey form for the baghouse filter.

If the dust is contained in process equipment other than a baghouse filter, use the Bursting Vessel Explosion Hazards Form.

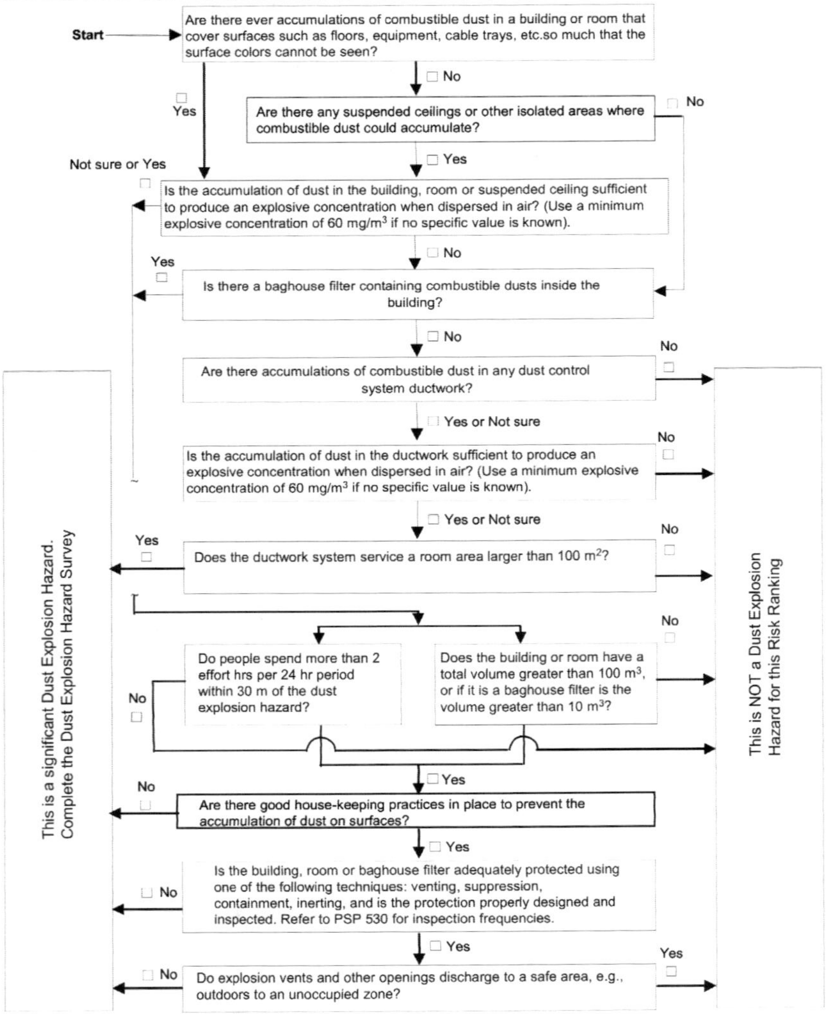

Figure 5.1 *Dust Explosion Hazard Selection.*

SECTION 6

BURSTING VESSEL EXPLOSION HAZARDS SURVEY FORM

Guidance:
Complete a Bursting Vessel Explosion Hazard survey form in section 6.2 for each vessel, tank, or other enclosure that might be a bursting vessel explosion hazard. This includes unfired pressure vessels, storage tanks, bins, silos, etc. Baghouse filters are covered in Section 5. Systems such as: Yankee dryers, designated high risk boilers, Sodium Percarbonate / Sodium Perborate, etc. are covered in Section 7 as Special Known Risks.

Guidance:
Fill out a bursting vessel survey form for each vessel, tank or enclosure that meets the pressure and volume criteria in Table 6.1, **and** then is selected by the flowchart in Figure 6.1.

6.1 STEPS TO FOLLOW

Follow steps (a), (b) and (c) below to determine if your vessels, tanks, and enclosures will be included in the survey.

(a) The following vessels are NOT bursting vessel explosion hazards for the purpose of this project:

1. Vessels with a volume less than 40 liters.
2. Vessels that normally operate completely filled with liquid, such as a liquid heat exchanger
3. Steam heat exchangers.

(b) Use Table 6.1 to identify vessels according to their pressure and volume criteria. The table is used to check if vessels can contain enough stored energy to be included in this survey. Enter Table 6.1 with the design pressure (MAP) of the vessel. Then, read across to the gas or vapor volume, and include vessels that exceed the volume for a given pressure. For example, if a vessel with an MAP of 10 bar gauge has a volume greater than 3 m^3, include this vessel in step (c). If a specific MAP is not listed in Table 6.1, use the nearest MAP listed in the table and interpolate to get a volume.

Table 6.1 *Bursting Vessel Pressure and Volume Criteria.*

	MAP		Gas or Vapor Volume		Typical Vessel Size	
	bar gauge	**kPa gauge**	**cu. meters**	**liters**	**Dia (m)**	**St Side (m)**
Non-Pressure Vessel	0.07	7	22500	22500000	24	48
	0.15	15	6200	6200000	16	32
	0.35	35	1100	1100000	9	18
	0.70	70	350	350000	6	12
	1.00	100	175	175000	4.5	10
Pressure Vessel	1	100	80.0	80000	4.0	6.5
	3	300	16.0	16000	2.5	3.3
	5	500	8.0	8000	1.9	2.8
	10	1000	3.0	3000	1.4	2.0
	15	1500	1.9	1900	1.1	2.0
	20	2000	1.3	1300	1.0	1.6
	25	2500	1.0	1000	0.9	1.5
	50	5000	0.4	400	0.7	1.1
	75	7500	0.25	250	0.6	0.9
	100	10000	0.2	200	0.5	0.9

(c) Use the flowchart Figure 6.1 as a second step to determine if a vessel should be included in the survey. The flowchart may exclude some vessels that meet the pressure and volume criteria. For example a typical P&G plant with an effective technical safety program might have just a few vessels that meet the bursting vessel explosion hazard criteria in the flowchart, and thus need a survey form. A large plant with many operations might have several vessels that need a survey form. A recent acquisition plant that does not have a well-established technical safety program may have more vessels that meet the criteria. Vessels in a well managed technical safety program should be designed, fabricated, protected, inspected and maintained properly.

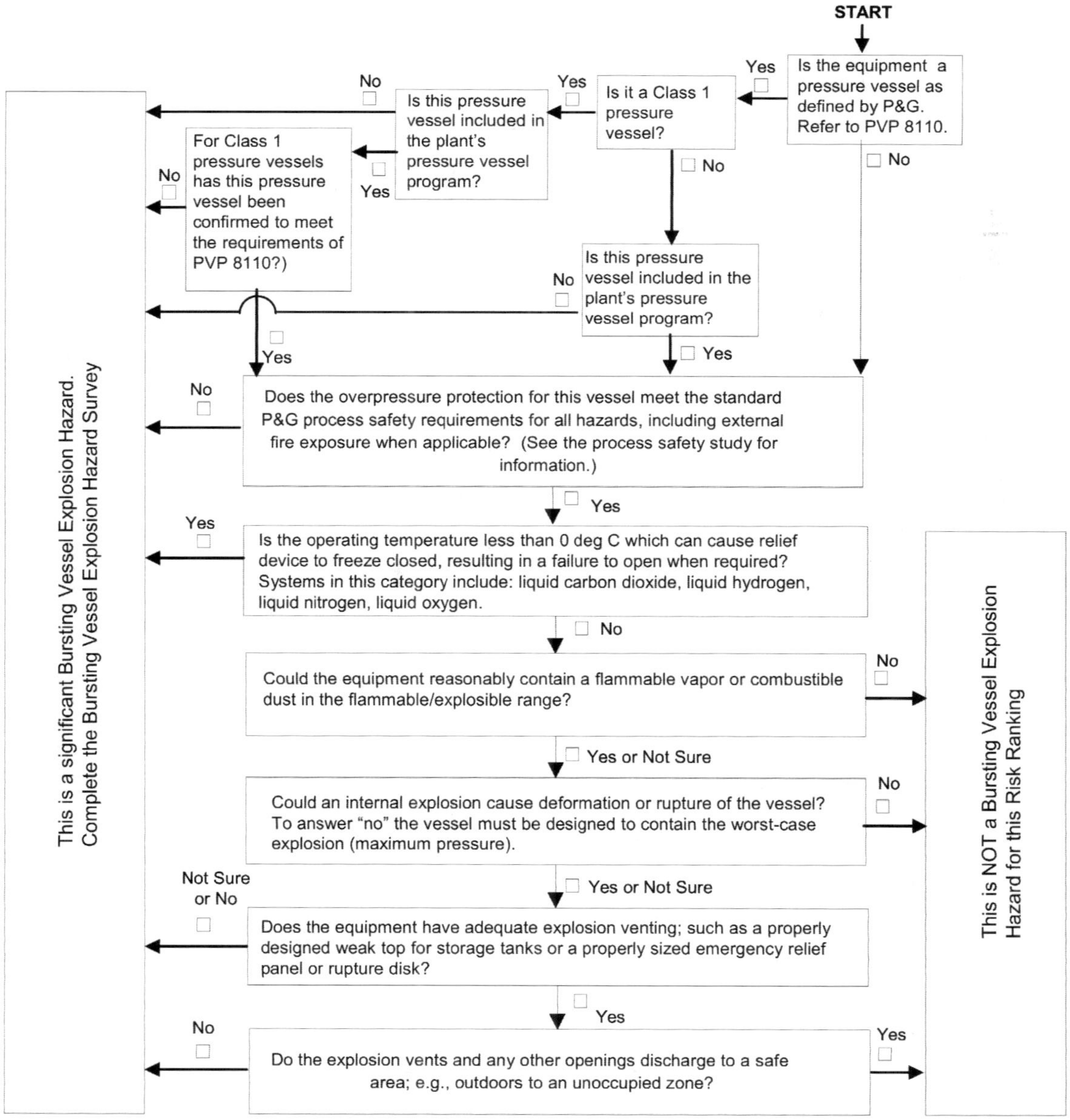

Figure 6.1 Bursting Vessel Flowchart

SECTION 8

SYSTEM LIKELIHOOD MODIFIERS SURVEY FORM

8.1 LIKELIHOOD MODIFIERS

Guidance:
When completing this survey form do not consider the entire site. Instead complete this survey form for each specific system identified in Section 4 (Chemical Hazard), 5 (Dust Explosion Hazard), 6 (Bursting Vessel Explosion Hazard) and 7 (Special Known Risk). Therefore if you had an ethanol storage system, ethanol processing system and 2 Yankee dryers on your sites, you would be expected to complete Section 8 four times (once for each system) and submit four Section 8 survey forms.

Indicate the name of the system to which the likelihood modifiers apply ________________________
Guidance:
If you are completing the electronic survey form, the Likelihood Modifiers from Section 8 are integrated into the survey forms for the various systems and therefore you do not need to indicate the system which the Likelihood Modifiers apply.

8.2 RESPONSIBILITY & ACCOUNTABILITY *(Check one box only that best describes your site)*

Guidance: Currently each Class 1 Chemical system must have an identified system owner. Ideally, every potential high risk system, which could lead to a Class IV Technical Safety incident would have a system owner. Currently there are many systems which do not require system owners under the current KE program, so if your system does not have a system owner (irrespective of whether the KE program requires one or not), check the "Less than Basic" box. Indicate which statement best describes your site:

❑ Less than Basic	❑ Basic Requirements	❑ More than Basic
• This system does not have a designated system owner.	• This system has a designated system owner. • The system owner is more than 80% trained and qualified on the system. • The system owner is responsible and accountable for managing the safety aspects of this system.	In addition to the base requirements: • The system owner is 100% trained and qualified on the system. • The system owner is knowledgeable of the process operation, maintenance, chemistry, and equipment. • The system owner is involved in things such as: reviewing changes to the system, qualification of new people, refresher training, review and update of procedures, major maintenance activities, and start up after major shutdown.

8.5 TESTING, INSPECTION, MAINTENANCE, REPAIR & MECHANICAL INTEGRITY *(Check one box only that best describes your site)*

Guidance: Testing and inspection include all Class 1 Chemical tanks and piping, process safety devices, combustion safeguards, thermal equipment, pressure vessels, electrical power safety devices and protective equipment, and welding.

❑ Less than Basic	❑ Basic Requirements	❑ More than Basic
• Less than 80% of the process safety devices, combustion safeguards, boilers and thermal equipment, PVs, tanks, piping, and electrical power equipment are identified and inspected and tested per laws, codes, and intent of Company practices. • Most tests, inspections and repairs are conducted by trained and qualified personnel. • Less than 80% of the inspection and test reports are complete, signed and dated. • Less than 80% of previous inspection and test reports are retained in a good data management system. • Less than 80% of testing and inspections are performed at the correct frequency aligned with laws, codes and the intent of company practices. • Less than 80% of failures, deviations and problems have been completed and closed-out. • Thorough checks are not always made to ensure that maintenance and repair materials, piping, spare parts, and equipment are suitable for the application in this system.	• More than 80% of the process safety devices, combustion safeguards, boilers and thermal equipment, PVs, tanks, piping, and electrical power equipment are identified and inspected and tested per laws, codes, and intent of Company practices. • All tests, inspections and repairs are conducted by trained and qualified personnel. • More than 80% of the inspection and test reports are complete, signed and dated. • 100% of testing and inspections and repairs are conducted following written procedures • More than 80% of previous inspection and test reports are retained in a good data management system. • More than 80% of testing and inspections are performed at the correct frequency aligned with laws, codes and the intent of company practices. • 100% of failures, deviations, and problems are identified, tracked. • More than 80% of failures, deviations and problems have been completed and closed-out. • There are written maintenance and inspection procedures for all unique or critical equipment.	In addition to the basic requirements: • 100% of process safety devices, combustion safeguards, boilers and thermal equipment, PVs, tanks, piping, and electrical power equipment are identified and inspected and tested per laws, codes, and intent of Company practices. • More than 95% of the inspection and test reports are complete, signed and dated. • More than 95% of previous inspection and test reports are retained in a good data management system. • More than 95% of testing and inspection are performed on time according to a schedule aligned with laws, codes and the intent of company practices. • More than 95% of recommendations are followed-up and completed. • Testing results provide actual test set points, thorough description of conditions found, and corrective action(s) taken. • All failures, deviations and problems are thoroughly followed-up in a timely manner. • Written safe practices cover procedures for performing all testing, maintenance, and repairs safely. These include the use of work permits for opening process equipment or piping and restoring the system to operating condition after inspection / repair; the removal of blanks and jumpers; and communication of completed work to affected employees. • JSAs or quick risk predications are regularly used when performing maintenance or repair work.

8.6 OPERATING & MAINTENANCE PROCEDURES *(Check one box only that best describes your site)*		
❑ Less than Basic	❑ Basic Requirements	❑ More than Basic

• Safe operating procedures (SOPs) cover most operating conditions.
• SOPs are not clearly written and do not reflect the work performed in the field.
• Less than 80% of SOPs are up-to-date.
• Safety operating procedures are not easily accessible.
• Less than 80% of maintenance procedures are clearly written.
	• Safe operating procedures (SOPs) are based on Job Safety Analysis, PHAs, and human factors analysis.
	• SOPs are complete, clearly written and match the current process.
	• SOPs are written in collaboration with affected employees.
	• More than 80% of SOPs are reviewed annually or whenever changes are made to the system.
	• SOPs are available 24/7.
	• SOPs include:
	  • initial start-up, start-up following a maintenance turnaround and start-up or recovery from an emergency shut-down;
	  • normal, manual and emergency operation;
	  • normal and emergency shut-downs.
	• SOPs include the precautions necessary to prevent exposure, covering administrative and engineering controls and personal protective equipment.
	• More than 80% of maintenance procedures are clearly written. These include safe line breaking, equipment opening, testing, inspection, maintenance, repair, and include the properties and hazards of the systems and chemicals used. Following any maintenance work they cover restoring the system to its operating condition, including removal of blanks, locks, and jumpers; system verification and testing; and communication of completed work to affected employees.
	• Contractor employers ensure that each contractor employee follows site procedures.
		In addition to the basic requirements:
		• More than 95% of SOPs are fully up to date to reflect the best practice performed by the operators in the field.
		• The SOPs include the results of deviations from operating limits and the steps required to avoid and/or correct deviations.
		• SOPs include a description of the safety systems and their functions to ensure all affected employees are aware of the safety systems.
		• SOPs are available for temporary operations (cleanouts, etc.) and clearly specify what must be done and who may start-up the system following an emergency shutdown.
		• There are procedures in place to control the ingress of people (maintenance, contractors, visitors or other support people) into the process area.
		• Operating and maintenance procedures are used for training and qualification of affected employees.

Overfill Protective Systems - Complex Problem, Simple Solution

Angela E. Summers, Ph.D., President, SIS-TECH Solutions, LP
12621 Featherwood Drive, Suite 120, Houston, TX 77034
asummers@sis-tech.com, 281-922-8324 (phone), 281-922-4362 (fax)

1 Abstract

Overfills have resulted in significant process safety incidents. Longford (Australia, 1998), Texas City (United States, 2005), and Buncefield (United Kingdom, 2005) can be traced to loss of level control leading to high level and ultimately to loss of containment. A tower at Longford and a fractionating column at Texas City were overfilled, allowing liquid to pass to downstream equipment that was not designed to receive it. The Buncefield incident occurred when a terminal tank was overfilled releasing hydrocarbons through its conservation vents.

The causes of overfill are easy to identify; however, the risk analysis is complicated by the combination of manual and automated actions often necessary to control level and to respond to abnormal level events. This paper provides a summary of the Longford, Texas City, and Buncefield incidents from an overfill perspective and highlights 5 common factors that contributed to making these incidents possible. Fortunately, while overfill can be a complex problem, the risk reduction strategy is surprisingly simple.

2 Introduction

Loss of level control has been a contributing cause in three of the worst industrial incidents in process industry history.

- The Esso Longford explosion (September 25 1998) in Australia resulted in 2 fatalities, 8 injuries, and A$1.3 billion in losses. Esso's natural gas supply to the state of Victoria for commercial and residential uses was severely affected for 2 weeks. Millions residents experienced cold showers and freezing nights for over 20 days (1).
- The BP Texas City explosion (March 23 2005) in the United States caused 15 fatalities and more than 170 injuries (2). Facility production was profoundly affected for months after the incident. Losses to BP are in excess of $1.6 billion (2).
- The Buncefield explosion (December 11 2005) in the United Kingdom (UK) injured 43 people and devastated the Hertfordshire Oil Storage Terminal, which was jointly owned by Total UK Ltd and Chevron Ltd (3). Residences and commercial buildings in the area were structurally damaged with some requiring demolition. The economic impact on regional businesses is estimated to be in the range of £130–170 million (3). Total losses may be as much as £1 billion (3,4).

These incidents involve three different industries located in three different countries. Esso processed natural gas for commercial and residential distribution in Longford, Victoria, Australia. BP fractionated raffinate, a mixture of hydrocarbons, for recycle within the refinery in Texas City, Texas, US. The Hertfordshire Oil Storage Terminal is part of the Buncefield depot, which is the 5th largest oil depot in the UK (3) and located in Hemel Hempstead north of London. Each incident propagated uniquely, arriving at its final outcome through different mechanisms. Yet, all suffered the same process deviation of high level and all resulted in devastating consequences.

This paper provides an overview of how each incident propagated from loss of level control to catastrophic event. It then discusses the following factors that contributed to these incidents:

1. Lack of hazard recognition
2. Underestimating the likelihood of overfill
3. Excessive reliance on operator
4. No defined safe fill limit
5. Inadequate mechanical integrity

Finally, it provides a simple 7 step solution for overfill protection.

3 Incident summaries

3.1 Esso Longford

The Longford incident occurred in the lean oil absorption unit that processed gas from the Bass Strait platform. Lean oil entered the top of an absorption tower and absorbed the heavier fractions (C2-C4) from the gas/condensate feed stream. Rich oil exited the tower just above the bottom section and was sent to downstream equipment, where the lean oil was recovered for recycle and the heavier fractions were collected for further processing. Condensate in the bottom section of the tower was recirculated through a reboiler to remove light ends.

The incident occurred when excess flow from the Marlin Gas Field (1) introduced more condensate into the absorption tower than it was designed to handle. Condensate overflowed the tower bottom section, mixed with the rich oil, and passed downstream to the rich/lean oil circulation system. The upset affected the demethanizer tower (15) and eventually caused high level in a separator drum initiating shutdown of the lean oil pump.

Without lean oil recirculation, the system chilled well below normal operating temperatures. The demethanizer reboiler (15) temperature went as low as -48C, causing cold temperature embrittlement (1, 14). The lean oil pump remained unavailable for more than 3 hours and when it was started up the hot lean oil caused the reboiler to stress fracture (1). The release resulted in a vapor cloud that spread outward for 170 meters prior to reaching fired equipment that provided an ignition source (1).

3.2 BP Texas City

The BP facility in Texas City, Texas, is one of the largest refineries in the United States with a capacity of 437,000 barrels per day (2). The raffinate splitter is part of the isomerization unit and it fractionated a mixture of hydrocarbons for internal recycle within the facility. The splitter was designed to relieve overpressures through a series of pressure relief valves (PRVs) that discharged into a blowdown drum. The drum was designed to trap mists and entrained liquids for discharge to the process sewer and to relieve vapors to the atmosphere.

During a cold start-up of the raffinate splitter, level is accumulated by starting the feed to the splitter with the outlet valve closed. When the level achieves a specified setpoint defined in the operating procedure, the operator controls the level by opening the outlet valve manually or by placing the level controller in automatic operation.

The incident occurred when the raffinate splitter was fed at normal rates for more than 3 hours (10) with the outlet closed during a cold start-up. Liquid overflowed the splitter into the vapor discharge header resulting in the opening of the PRVs. Liquid surged through the PRVs into the blowdown drum and began draining into the process sewer. Within seconds, the flow from the splitter overwhelmed the drain capacity yielding a geiser of hydrocarbons from the drum stack that rained down inside and outside the process area. The hydrocarbon vapor and liquids eventually reached an ignition source resulting in the catastrophic explosion.

3.3 Buncefield UK

The Hertfordshire Oil Storage Terminal is part of a complex of tank terminals known as the Buncefield Depot. The depot has an estimated capacity of 60 million gallons and serves as a major distribution center for the UK oil pipeline network (5). It provides fuel to Humberside, Merseyside, as well as to Heathrow and Gatwick airports (3).

The incident occurred when the automated tank gauging system failed allowing excess fuel to be fed into a terminal tank for 11 hours (11). The fuel overflowed through the tank conservation vents for approximately 40 minutes (13) prior to ignition, producing a large vapor cloud estimated to be 8 hectares in size (12). At the time of the release, ambient conditions were near freezing with a dense fog. When the vapor cloud ignited, the resulting explosion caused significantly more damage than consequence models predicted. The cold, icy, and wet ambient conditions are believed to have intensified the explosion, which is currently considered the largest peacetime explosion in European history (11) producing a tremor measuring 2.4 on the Richter scale and blowing out windows five miles away from the site (12).

4 Five gaps

4.1 Lack of hazard recognition

In the majority of processes, level has little significance to plant production or product quality. The absolute level often varies over a large range where the "normal" operating level is not well-defined or tightly controlled. In the BP and Esso incidents, the normal operating level was significantly below what would threaten the column integrity. At Buncefield, the operating level was simply inventory to be managed and normally varied across a large range.

High level is often not a hazard itself. Instead, the hazard is too much mass or volume. Some overfills challenge the tank or vessel where the level is accumulating, causing it to overpressure or to collapse when the retained mass exceeds its structural design limits. Many overfills result in loss of containment when liquid passes to downstream equipment that is not designed to receive it.

At Texas City, the hazard analysis of the raffinate splitter did not consider high level in the splitter to be credible, because it would take more than 3 hours for the splitter to overfill (10). However, lifted column trays are frequently found during internal inspections of columns in the process industry, indicating that high level does occur in many columns. Experience has also shown that sometimes operators deliberately retain liquid in columns to ride out upsets of downstream equipment.

No hazard analysis was performed at Longford, but it is likely that the team would have assumed that the absorption tower was designed to receive whatever liquids came to it. The likelihood of overfill worsens as debottlenecking allows increased production rates that yield a higher rate of level rise than historical experience indicates. At Longford, the tower rapidly overfilled due to excess flow from the platform (1).

Overfill hazards vary depending on the type of vessel and associated upstream and downstream equipment. When interconnected equipment is affected, the hazard analysis should ensure that the high level hazard is prevented and is not allowed to pass downstream. It is rarely effective to allow a high level event to propagate and to depend on downstream process variables to be fast enough to prevent equipment damage. For example, high level in a knock-out drum requires immediate response to protect the compressor from damage. Waiting until high compressor vibration is detected is too late.

4.2 Underestimating the likelihood of overfill

Level seems so simple to detect that anyone should be able to recognize it and respond in a timely manner. Unfortunately, high level can rarely be seen directly by the operator. It is just one of many process variables on the display. Compounding this perception, level often does not affect the unit operation or cause any other significant process variable disturbance until the safe fill level is exceeded and suddenly the mechanical integrity of the vessel or interconnected equipment is threatened.

High level may have different causes in each mode of operation, e.g., start-up, normal, or upset conditions. Start-up may require the accumulation of level, so the outlet control valve is initially in manual operation and closed until the normal operating level is reached. Level may vary over a large range during normal operation. During upsets, operators may operate vessels at higher than normal levels to smooth out process operation by using the available capacity as a dampener for upsets in upstream or downstream equipment.

Some hazard analysis teams erroneously believe that overfill is not a credible event, because the time required to fill the vessel is generally on the order of minutes or hours rather than seconds. Some events propagate slowly, such as the rise of level in a product storage tank, while others occur quickly through a random event, such as a process upset sending excess liquid to a knockout drum for a compressor. The slower the event the greater the tendency to believe that the operator can adequately address the event; likewise, the more sporadic the event, the greater the tendency to believe the event will not last long enough to cause overfill. Believing that high level is non-credible is especially attractive when the existing design has no provision for high level alarm or trip.

Estimating the likelihood of overfill is complicated by the combination of manual and automated control that is necessary as the equipment is started up and operated. Figure 1 shows the range of automation commonly found in tank farms and terminals. The degree of automation is generally related to the expected rate of level rise and operator work load. Automated control and safety systems are generally added when control changes must be made too often to be continuously managed by the operator or when work complexity has increased to the point where the expected human error rate is no longer acceptable.

A safe operating limit must be specified and the consequence of exceeding it should be explained in the operating procedures. Without clearly stated limits and consequences, the operator will not adequately monitor level, especially during intense work periods. Overfill is a credible event and it takes good operator procedures to reduce its likelihood.

4.3 Excessive reliance on operator

In the Esso, BP, and Buncefield events, the initial incident investigation blamed the operator for not maintaining level. BP took disciplinary action against supervisors and hourly employees who were directly responsible for operating the isomerisation unit on March 22 and 23, blaming operators for ignoring procedures (6). Esso blamed the control room operator for ignoring alarms and not following procedures even though the system generated more than 300 alarms per day (7, 8).

The "blame the operator" tendency is encouraged by the length of time required to reach overfill. In many applications, the operating basis provides adequate time for the operator to control the level within acceptable tolerance, but human error is always possible. Work load and piping network complexity decrease the operator's ability to reliably control level and maintain process safety. Some facilities have been debottlenecked and expanded to increase production. This has increased operator work load and eroded the time available for operator response to abnormal events. In some cases, the available time has been reduced to the point where manual response is no longer effective and automatic level control must be considered.

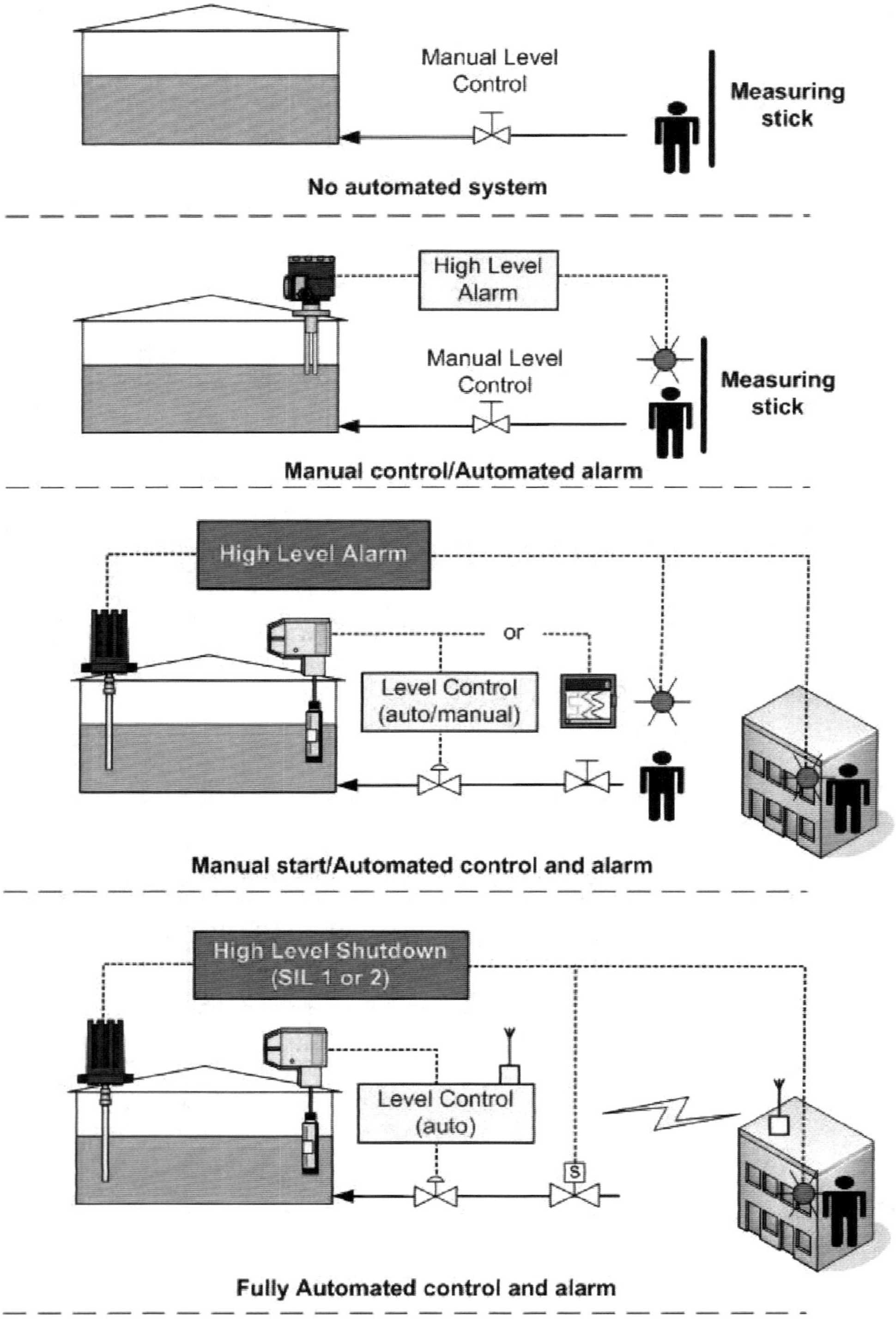

Figure 1. Examples of automation commonly found in tank farms and terminals

Personnel hazards should be considered when directing operators to take manual actions in the process unit in response to high level, such as draining knock-out drums. Local response generally moves the operator into the hazard zone increasing the risk to that individual. Consequently, the design must provide sufficient time for the operator to take response and means to verify the process response. Further, there should be time to evacuate the area if the process response is not as expected. When fast response is required, operator drills should be considered to allow the operator to practice the response and to verify the time required to detect and respond. These drills can identify issues with the design, installation, and labelling, as well as with the procedures and training.

Automated controls are often added to increase operating efficiency and reliability. They should also be considered to reduce operator exposure to the hazardous event. For significant hazardous events, automated trips ensure protection is continuously provided even when the operator is focused on other duties. An automatic trip can detect high level and prevent filling beyond the safe fill limit.

4.4 No defined safe fill limit

In many applications, the entire level range from empty to postulated failure point is not displayed. Instead, the expected operating range is covered by the measurement device. This provides the most accurate measurement across the operating range, while unfortunately leaving the operator with no indication of the level when it rises above the normal operating range.

For example, in the BP Texas City design, the column level was measured in the first ten feet of the column only. When the transmitter range was exceeded, the level on the operator display remained unchanged even though the level continued to rise. The high level alarm was also located in the bottom part of the column, indicating an abnormal process condition, but not an unsafe one. It was considered normal during start-up for the column to be filled higher than the alarm setpoint. The unit had experienced many start-ups with high bottoms level without incident. The problem was that no limit was set on how high is unsafe.

A safe fill limit should be clearly established in the design basis. The safe fill limit is specified based on an understanding of the postulated failure level, the analytical capability of the instrumentation used for the measurement, the fill rate, and the time required to achieve a safe state. The safe fill limit should ensure that action can be completed prior to reaching the postulated failure level. It should be conservatively estimated based on expected measurement drift in the process and environmental conditions.

Figure 2 shows the transition of the level from the normal operating range to the postulated failure point. An alert may be provided to support level control and its setpoint should allow enough time for the operator to take response to prevent the level from reaching the safety alarm or trip setpoints. The safety alarm should provide enough time for the operator to bring the level back under control or to take the equipment to the safe state.

The trip point is selected to automatically initiate a feed shutdown so that the level rise is stopped prior to reaching the postulated failure level. The off-set between the trip setpoint and the safe fill limit is the design safety margin. When an alarm is also implemented, the alarm setpoint should be conservatively set below any trip setpoint, allowing the operator sufficient time to take the process to a safe state prior to the trip being initiated. The operator must be capable of taking effective action after receiving the alarm; otherwise it will be treated as a trip notification rather than something requiring action.

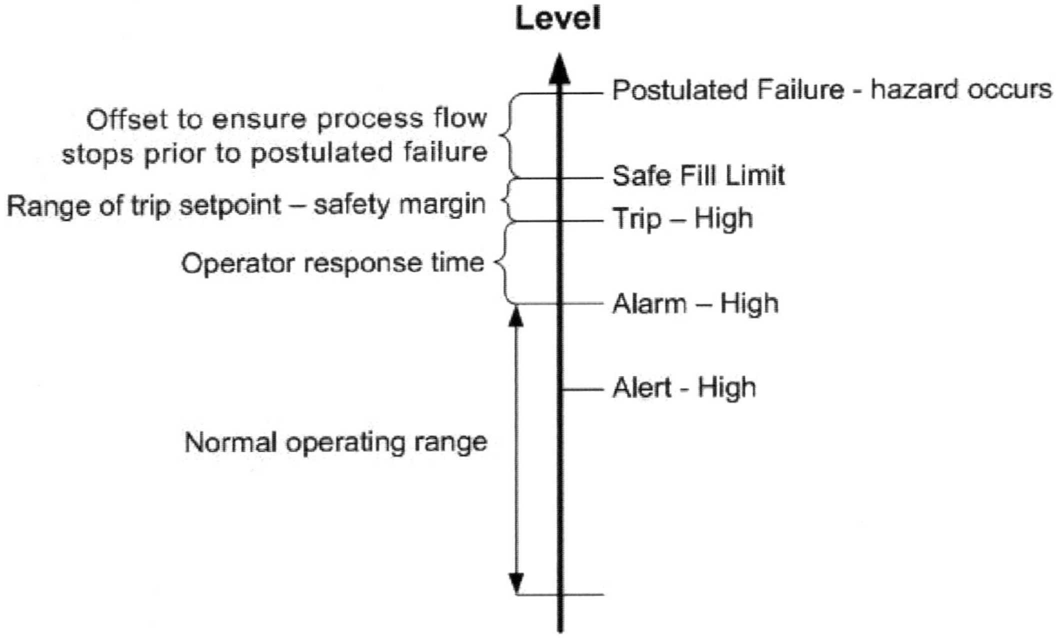

Figure 2. Range of Level Showing Transition from Normal Operating Range to Vessel Failure for High Level Events

4.5 Inadequate mechanical integrity

The Buncefield and Texas City incidents involved level switches that failed to operate correctly during the incident. This seems to support the concept that level switches are bad. However, evidence shows that the sensor was not installed properly in Buncefield and was not maintained in Texas City. No signal processor – digital or analog - works if the sensor is not installed properly or maintained. There are no bad level devices, only technology misapplications, improper installations, and inadequate mechanical integrity programs.

Some companies have mandated that only transmitters be used in safety services, stating that direct mounted switches should not be used due to their lack of continuous signal. This may be good practice for other process variables, but level is different. For columns and storage tanks, the safe operating limit is significantly outside the normal operating level, resulting in the high level alarm or trip sensors being at a very low output for long periods of time. Under this condition, the benefit of a diverse technology sensor, like a switch, may outweigh the advantages provided by a continuous signal. Consequently, it is an acceptable practice to implement storage tanks with an automated tank gauging system that uses an analog measurement covering the expected normal operating range and a discrete level switch that is used to initiate feed shutdown.

A properly maintained level switch can provide years of cost effective and satisfactory service. On the other hand, years of neglect will allow the most expensive device to fail. There are many technologies available for level measurement and detection, from simple float type discrete switches to new advanced guided wave radar transmitters. Each technology has characteristics that make it the right choice for a particular application (9). For most safety applications, the main considerations for equipment selection are the process operating mode, the operating environment, the historical equipment performance, and the ease of maintenance and testing.

A high level alarm and trip can be implemented with separate level switches at the selected points on the vessel or with a transmitter that covers both setpoints. Although transmitters may not increase diagnostics in services that normally do not have level, they do provide the ability monitor and alarm over a chosen range.

No matter what technology is selected; the mechanical integrity of the equipment must be maintained throughout its installed life. Functionality is demonstrated by forcing the sensor to “see the process variable” and to generate the correct signal at the specified setpoint. Testing must prove that the equipment can operate as required to prevent overfill. Although diagnostics can detect many types of failures, a proof test is necessary to demonstrate operation at the required setpoint. This is the only test that proves that the device works as required.

Buncefield had no provisions for full functional testing of the sensor without raising the level to the trip setpoint. Raising the level is a simple way to demonstrate device operation, but may present unacceptable risk in some applications. At Buncefield, no full test was ever performed, so the flawed installation was never corrected. All safety equipment must be installed with means that allow positive confirmation of their operation at the specified setpoint. This may include stilling wells or float chambers with facilities to simulate level.

5 Solution

Catastrophic overfills are easily preventable. When overfill can lead to a fatality, follow these 7 simple steps to provide overfill protection:

1. Acknowledge that overfill of any vessel is credible regardless of the time required to overfill.

2. Identify each high level hazard and address the risk in the unit where it is caused rather than allowing it to propagate to downstream equipment.

3. Determine the safe fill limit based on the mechanical limits of the process or vessel, the measurement error, the maximum fill rate, and time required to complete action that stops filling.

4. When operator response can be effective, provide an independent high level alarm at a setpoint that provides sufficient time for the operator to bring the level back into the normal operating range prior to reaching a trip setpoint.

5. When the overfill leads to the release of highly hazardous chemicals or to significant equipment damage, design and implement an overfill protection system that provides an automated trip at a setpoint that allows sufficient time for the action to be completed safely. Risk analysis, such as Layers of Protection Analysis (LOPA), should be used to determine the safety integrity level (SIL) required to ensure that the overfill risk is adequately addressed. While there are exceptions, the majority of overfill protection systems are designed and managed to achieve SIL 1 or SIL 2.

6. Determine the technology most appropriate for detecting level during abnormal operation. The most appropriate technology may be different from the one applied for level control and custody transfer.

7. Finally, provide means to fully proof test any manual or automated overfill protective systems to demonstrate the ability to detect level at the high setpoint and to take action on the process in a timely manner.

References

1. Wikipedia, "1998 Esso Longford gas explosion," http://en.wikipedia.org/wiki/1998_Esso_Longford_gas_explosion.
2. Wikipedia, "Texas City Refinery (BP)," http://en.wikipedia.org/wiki/Texas_City_Refinery_(BP).
3. Buncefield Major Incident Investigation Board, "The Buncefield Incident 11 December 2005: The final report of the Major Incident Investigation Board," ISBN 978 0 7176 6270 8, (2008).
4. British Broadcasting Station, "Buncefield blast could cost £1 bn," December 11, 2008, United Kingdom, http://news.bbc.co.uk/2/hi/uk_news/england/7777539.stm.
5. Wikipedia, "2005 Hertfordshire Oil Storage Terminal Fire," http://en.wikipedia.org/wiki/2005_Hertfordshire_Oil_Storage_Terminal_fire.
6. The Chronicle, Anne Belli, "BP blames staff for blast," May 18, 2005, Houston, Texas, http://www.chron.com/disp/story.mpl/special/05/blast/3186965.html.
7. Doig, Meredith, "Longford – Why efficiency is the enemy of safety," originally published in The Manager Online Magazine (now ceased) in June 1999, http://www.unisa.edu.au/corpsocialresp/casestudies/longford.asp.
8. O'Brien, Kerry, "Esso has cover-up strategy: Longford disaster allegations," http://www.abc.net.au/7.30/stories/s23848.htm.
9. Boyes, Walt, "First the application, then the product," ControlGlobal.com, http://www.controlglobal.com/articles/2007/022.html.
10. Mogford, J., "Fatal Accident Investigation Report: Isomerization Unit Explosion," Final Report, Texas City, Texas, USA (2005).
11. British Broadcasting Station, "Buncefield tank was overflowing," May 9, 2006, United Kingdom, http://news.bbc.co.uk/2/hi/uk_news/4752819.stm.

12. The Guardian, David Fickling, "Faulty fuel gauge caused Buncefield Explosion," United Kingdom, http://www.guardian.co.uk/uk/2006/may/09/buncefield.davidfickling.
13. The Daily Mail, "Petrol tank 'overflowing' before Buncefield blast," May 9, 2006, United Kingdom, http://www.dailymail.co.uk/news/article-385607/Petrol-tank-overflowing-Buncefield-blast.html.
14. Kletz, Trevor, "Lessons from Longford – The Esso Gas Plant Explosion," Chemical Engineering Progress, September 1, 2001.
15. Longford Gas Plant Accident and Victorian Gas Supply Crisis, Australian Government, Attorney General's Department, Emergency Management Australia, http://www.ema.gov.au/ema/emadisasters.nsf/6a1bf6b4b60f6f05ca256d1200179a5b/51561bea38d1b17cca256fbe007bb49b?OpenDocument&Highlight=0,esso

PRESSURIZATION OF FIXED ROOF STORAGE TANKS DUE TO EXTERNAL FIRES

Fabien Fouillen, INERIS, Parc Technologique ALATA 60 550 Verneuil-en-Halatte, France, *fabien.fouillen@ineris.fr, +33.(0)3.44.55.65.86*
Stéphane Duplantier, INERIS, Parc Technologique ALATA 60 550 Verneuil-en-Halatte, France, stephane.duplantier@ineris.fr, +33.(0)3.44.55.61.09

INTRODUCTION

This study deals with one of the phenomena which is able to generate fire balls like the one which was observed during the 1987 Edouard Herriot port of Lyon fire. Reflections led on this accident have pushed to consider the phenomenon of tank pressurization as a potential initiating event of the fire ball observed. In concrete terms, when a fixed roof storage tank is surrounded by a pool fire, and when the vents or the insulation are not correctly designed, the inside pressure can raise until the mechanical limits of the structure are reached. The aim of this study is to develop a tool which will allow to follow the variations of the pressure and temperature in the gas phase, highlighting if a pressure increase is possible or not. This will help the sizing of new tank vents and insulation of the walls, or evaluating the risk of already build structures, by solving the question occurrence of the accident, regarding the estimated external fire duration. The model also gives the temperature distribution in the liquid phase, at each time step. In case of rupture, this information will be useful to describe the phenomenological aspect of the accident after the rupture of the tank. It means that qualifying and quantifying the fire ball generated and its effect on persons and structures will be possible. Currently, although a few analytical models exist to quantify the fire ball and its effects after the rupture, none is able to describe with accuracy the pre-rupture phenomena. This model is a work in progress, and at the end, the tool will take into account the two phases of the accident and will allow to study complete and consistent scenarios. After a description of the numerical model developed, this paper presents its application on the study of the influence of various parameters.

DESCRIPTION OF THE MODEL AND INPUT DATA

The modelling tool presented is a dynamic model which represents the axisymmetric behavior over time of a cylindrical tank, when this one is impacted by a pool fire. The approach which has been followed in all building steps of the model was to "forget" the empiric assumptions, laws or formulas used in current standards, and to consider only the basic equations of thermodynamic and fluids mechanic. As the literature doesn't present experimental tests or accidents analysis allowing to check the evolution of different parameters, it is essentially the study of energy balance and trends evolutions which have been considered to validate each step of conception of the model. The heat transfers taken into account in the model are represented on Figure 1.

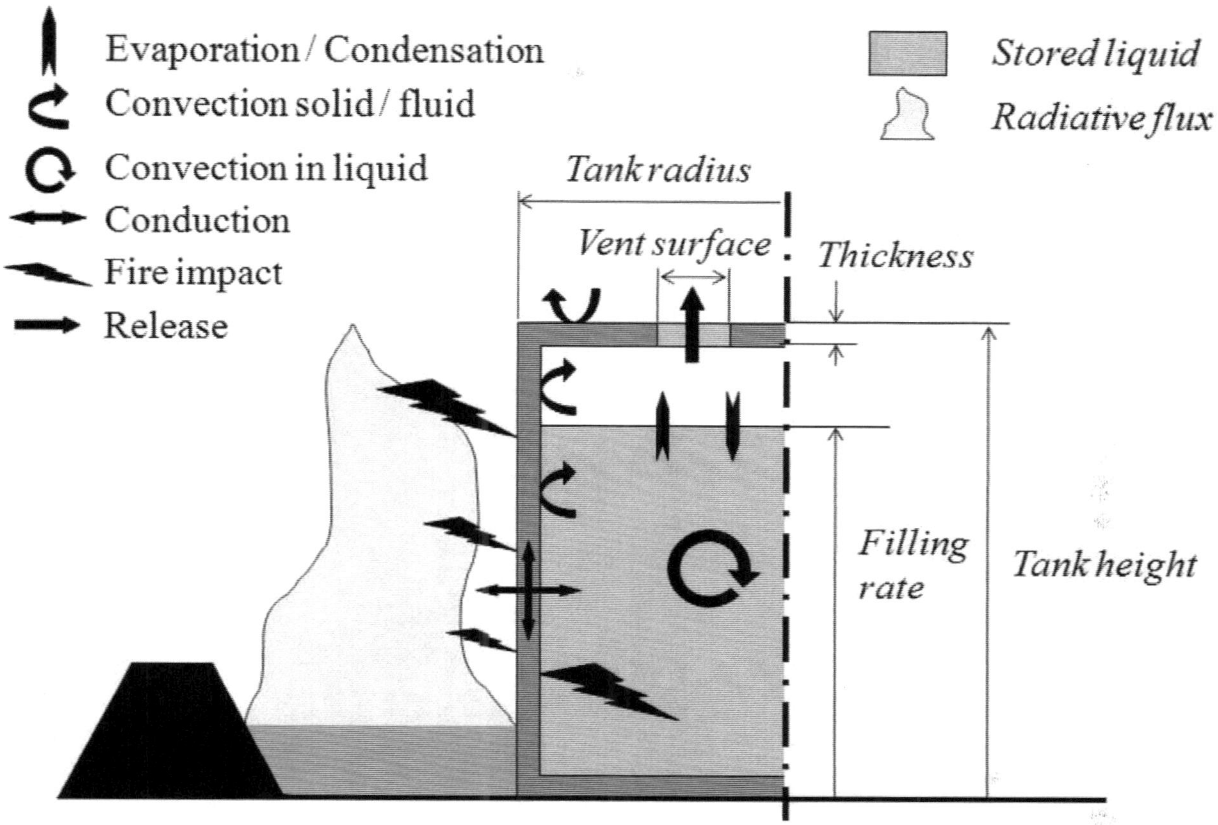

Figure 1. Input data and considered heat transfers

As stated previously, the studied geometry is considered axisymmetric. The input parameters written in italic on Figure 1, are the specific dimensions of the tank: its height, diameter and thickness. The thickness of the wall is supposed uniform. It's also necessary to input the filling rate of the tank, expressed as a percentage. The vent itself is defined by its surface and its discharge coefficient. A minimum opening pressure must also be provided. This latter parameter allows to take into account the safety valves. It is possible to consider several substances: water, ethanol, hexane (C_6), decane (C_{10}), undecane (C_{11}) or dodecane (C_{12}). In all cases, the substance is considered pure. The characteristics of the substances are intrinsic functions of temperature. To model the fire that impacts the tank, the radiative flux per unit area of flame must be defined. It is possible to define two types of aggression: a uniform distribution of the radiative flux along the entire height of the flame, or a radiative flux function of the height. In all cases, the flame height is limited to the height of the tank.

MESH

The mesh put in place is axisymmetric. The volume of liquid is divided into m x n identical sub-volumes. There are m volumes along the height, and n following the radius of the tank. So as to work with volumes of the same capacity, the spatial distribution is done following the equations **(1)** and **(2)** :

$$\Delta z = \frac{H_l}{m} \quad \textbf{(1)}$$

$$\Delta x = \frac{D}{2}\left(\sqrt{\frac{n-i+1}{n}} - \sqrt{\frac{n-i}{n}}\right) \quad \textbf{(2)}$$

The gas is supposed to have homogeneous pressure and temperature on the entire free volume. Therefore, it is composed of a single computational cell. As for the walls of the tank, it is represented with m+1 cells along the height, with a width along the radius equal to the thickness of the wall, as shown in Figure 2.

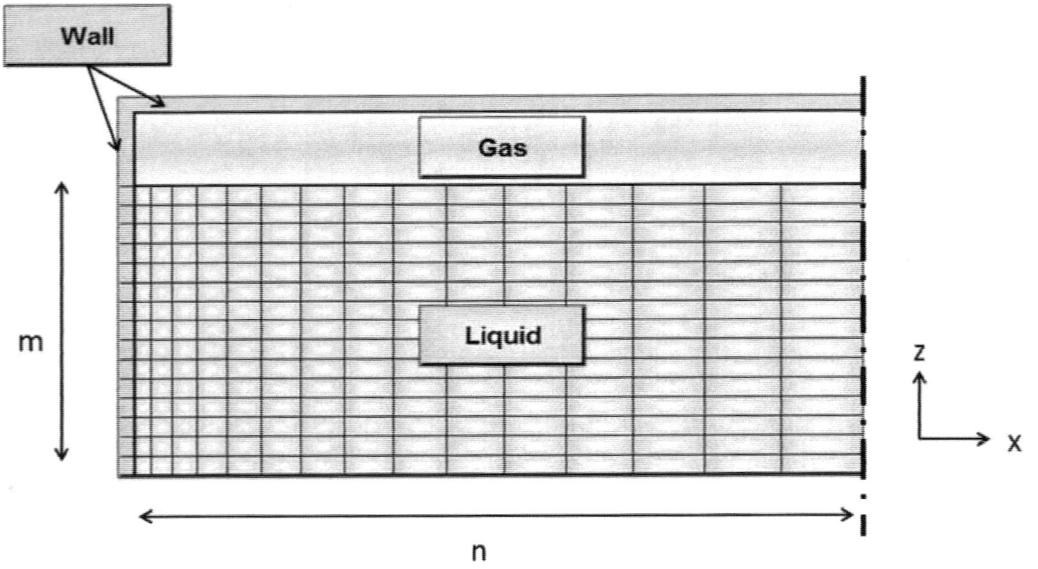

Figure 2. Axisymmetric mesh used

HEAT TRANSFERS

Thermodynamics allows us to predict the total amount of energy that a system should share with the outside to move from one equilibrium state to another. Thermokinetic is used to describe quantitatively the changes in characteristics of the system, particularly temperature, between the initial state of equilibrium and final steady state. At each time step, the calculation of heat transfers leading to a variation of pressure in the gas phase is done. The Figure 3 shows which equations are used in each cell so as to equilibrate the energy balance. The numbers in this Figure refers to the numbering of equations which are all defined later in the paper.

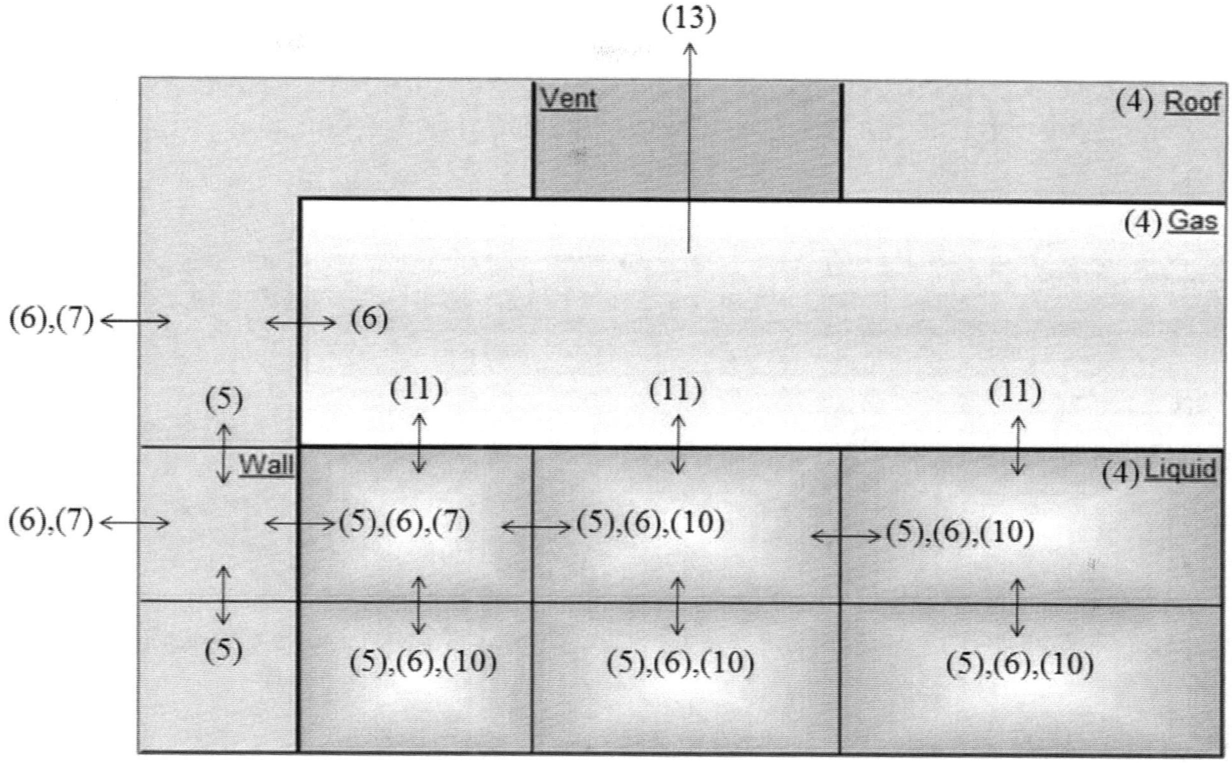

Figure 3. Equations taken into account

The various equations used correspond to the first thermodynamic law applied to each defined system, i.e. to the following energy balance:

$$\varphi_i + \varphi_{ge} = \varphi_o + \varphi_{st} \quad \textbf{(3)}$$

With φ_i the incoming heat flux, φ_{ge} the generated heat flux, φ_o the outgoing heat flux, φ_{st} the stocked heat flux. Solving this equation leads to a differential equation which allows to determine the evolution of the temperature at a given point during time. The heat flux are calculated the following ways:

- ***The generated heat flux:*** it occurs when another form of energy (chemical, electrical, mechanical, nuclear) is converted into thermal energy. The system considered here is not concerned by this type of flux.
- ***The stocked heat flux:*** it is linked to an augmentation of the internal energy of the system during time. It can be written as **(4)**:

$$\varphi_{st} = \rho V C \frac{\partial T}{\partial t} \quad \textbf{(4)}$$

- ***The incoming and outgoing heat flux:*** heat transfers can occur by:

Conduction: it is the favorite way of energy transfer in opaque medium. It occurs without material movement, along the temperature gradient. It is calculated following the Fourier hypothesis (for example in the x direction) by the equation **(5)**:

$$\varphi = -\lambda S \frac{\partial T}{\partial x} \quad \textbf{(5)}$$

Convection: it is the transfer of heat between a solid and a fluid, obtained by the displacement of the fluid. This phenomenon is governed by the Newton law **(6)**:

$$\varphi = hS(T_P - T_\infty) \quad \textbf{(6)}$$

Radiation: it is an electromagnetic transfer of energy between two surfaces. It is calculated with the equation **(7)**:

$$\varphi = \sigma \varepsilon_P S\left(T_P^4 - T_\infty^4\right) \quad \textbf{(7)}$$

Evaporation, condensation and convection inside the liquid are important heat transfer modes. Taking them into account is more complex their modeling is detailed in the next sections.

CONVECTION ALGORITHM INSIDE THE LIQUID

The convection algorithm developed is based on the use of the dimensionless Grashof's number. A static fluid in contact with a wall brought from a temperature *T* to a higher temperature *T+ΔT*, will overheat by conduction. The density of a fluid volume will increase from ρ to $\rho+\Delta\rho$. This unit volume will be subjected to a lifting force. While β is the cubic expansion coefficient, $\beta\Delta T$ represents the module of the acceleration produced by thermal expansion due to changes in temperature. This movement of fluid caused by differences in density will give rise to convection currents. By equalizing the variation of kinetic energy to the variation of potential energy, the number of Grashof allows to calculate the rising speed (ms^{-1}), for a unit volume of liquid:

$$Gr = g\frac{\Delta\rho}{\rho}\frac{L^3}{\mu^2} \quad \textbf{(8)}$$

$$u = \sqrt{2Gr}\frac{\mu}{\rho L} \quad \textbf{(9)}$$

This algorithm is based on the principle that the phenomenon of convection is initiated and amplified by the volume of liquid moving upward. The downward movements are the result of the imbalance caused. According to equation **(9)**, the fraction of volume which comes out of a cell is defined by:

$$dV = S_h\, dz = S_h\, u\, dt \quad \textbf{(10)}$$

Mass conservation law is used to define step by step, for each cell, the fraction of liquid volume entering or leaving the considered cell, as well as its origin or its direction. The contribution of convection for calculating the temperature of a fluid cell density can be determined following the approach detailed in Figure 4.

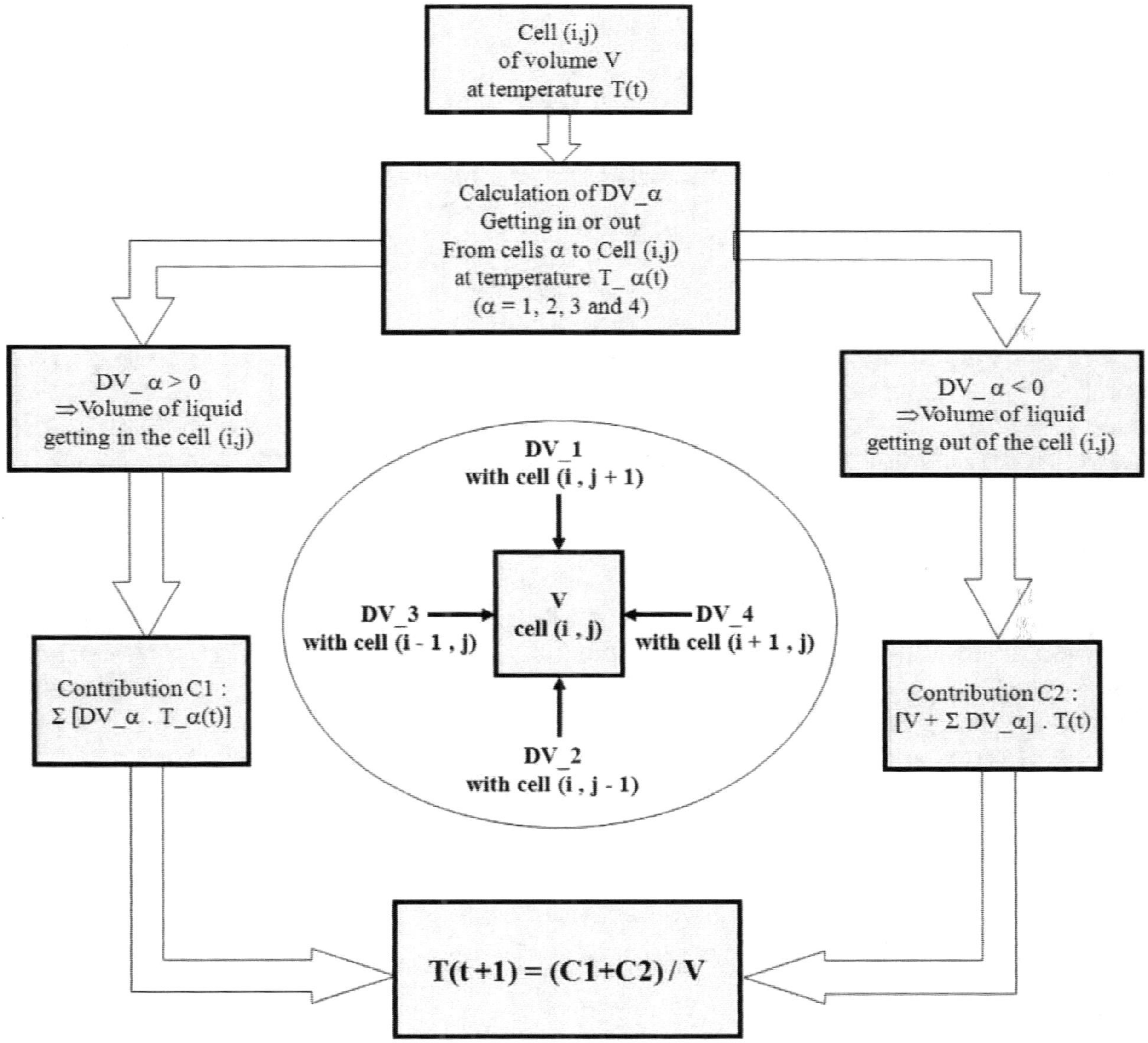

Figure 4. Participation of natural convection in liquid temperature calculation

INTEGRATION OF EVAPORATION AND CONDENSATION

At each time step, and for each cell volume, the vapor pressure at the liquid temperature is compared to the partial vapor pressure in the gaseous volume of the tank. This allows to evaluate the mass transfer between phases in order to determine the contribution of evaporation and condensation to variations of:

- Pressure;
- Liquid level;
- Temperature.

The total number of moles of liquid which have to be evaporated is calculated so that the balance is achieved between the liquid and gas phases. To do so, the equation **(11)** is used. In this formula, *B* represents the second virial coefficient.

$$N(i,j)=\frac{\left(P_{vap\_sat}-P_{partial\_vap}\right)V_g}{RT_g\left(1+\dfrac{B}{V_g}\right)} \quad \textbf{(11)}$$

Two cases may arise: either the cell (i, j) is likely to contribute to the change of mole number in the gas phase, and the algebraic value of *N* is calculated, or it isn't, and *N* is equal to zero. This approach prevents from the presence of liquid with temperature above the equilibrium temperature corresponding to the partial pressure in the gas.

TAKING THE VENT INTO ACCOUNT

The gas is composed of a mixture of air and vapor due to the evaporation of the liquid inside the tank. The pressure in the tank is a function of four interdependent parameters: the temperature of the gas mixture, the rate of evaporation/condensation, the flow rate of the vent, the liquid level. At each time step, evaporation contributes to the change in the partial pressure of vapor, and increasing the total pressure in the gas. When it reaches and exceeds the trigger pressure of the vent, a certain amount of gas is evacuated outside of the tank. This volume of matter is calculated with the Bernoulli's equation:

$$V_{evac}=2C_D S_{vent}\sqrt{\frac{\Delta P}{\rho}}\,\Delta t \quad \textbf{(12)}$$

With C_D the discharge coefficient of the vent, S_{vent} its surface and ΔP the differential pressure between the inside and the outside of the tank. Knowing the number of moles of liquid at a temperature $T_l(i,j)$ evaporated, and the number of moles of gas at temperature T_g evacuated by condensation or through the vent, it is possible to determine the contribution of gas transfers to the variations of gas temperature:

$$T_g(t+1)=\frac{\left(N_{mol\_vap}+N_{mol\_air}+N_{mol\_cond}-N_{mol\_évac\_vap}-N_{mol\_évac\_air}\right)T_g(t)+\sum N_{mol}(i,j)T_l(i,j)}{N_{mol\_vap}+N_{mol\_air}+N_{mol\_cond}-N_{mol\_évac\_vap}-N_{mol\_évac\_air}+\sum N_{mol}(i,j)} \quad \textbf{(13)}$$

INFLUENCE OF THE DIFFERENT PARAMETERS

So as to determine the influence of the accessible parameters on the behavior of the system, a reference case has been used. The parameters corresponding to this reference case are:

- The height of the tank: 14 m;
- The diameter of the tank: 24 m;
- The thickness of the walls: 10 mm;
- The filling ratio: 50% of the tank capacity;
- The substance: hexane;
- The vent surface: 0,1 m^2;
- The maximal radiative flux of the fire: 50 $kW.m^{-2}$.

The variations of the pressure of the gas phase are presented in Figure 5. The maximum pressure obtained here is called Pmax_ref. Pmax_ref keeps the same value all along the present section.

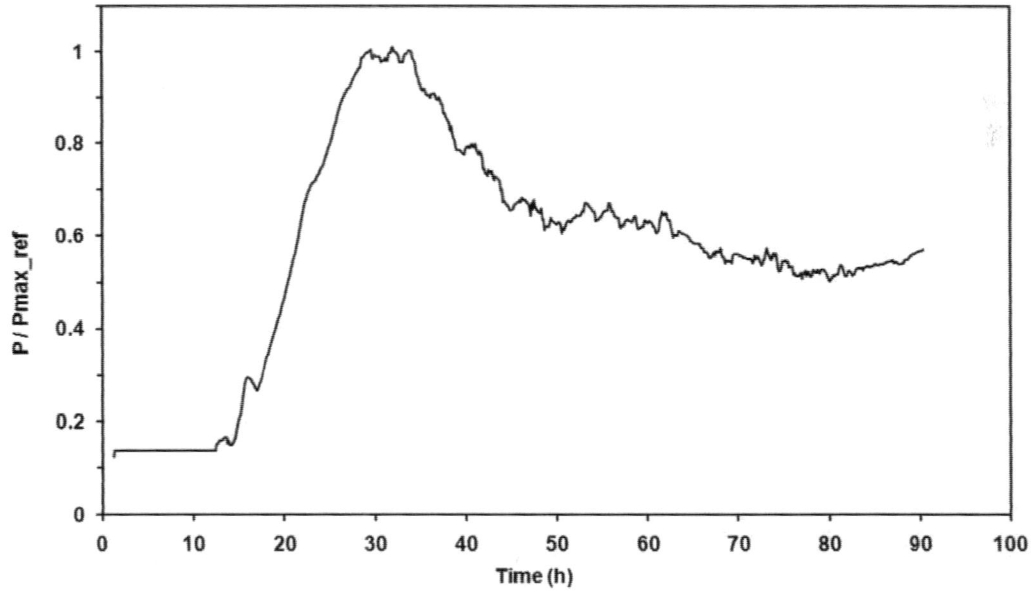

Figure 5. Evolution of pressure for the reference case

During the first 15 h, the size of the vent is sufficient to compensate the vaporization rate. The maximal overpressure is observed after 30 h of fire, where a 5 h plateau is reached. After, the pressure starts to decrease with the level of the liquid.

The influence of the followings parameters has been evaluated:

- The stored substance;
- The liquid level;
- The diameter of the tank;
- The vent surface.

The detailed graphs of the results obtained are presented from Figure 6 to Figure 9. They show the variation of the pressure in the gas phase when only one of the parameters previously defined varies. The blue curve is always the reference case.

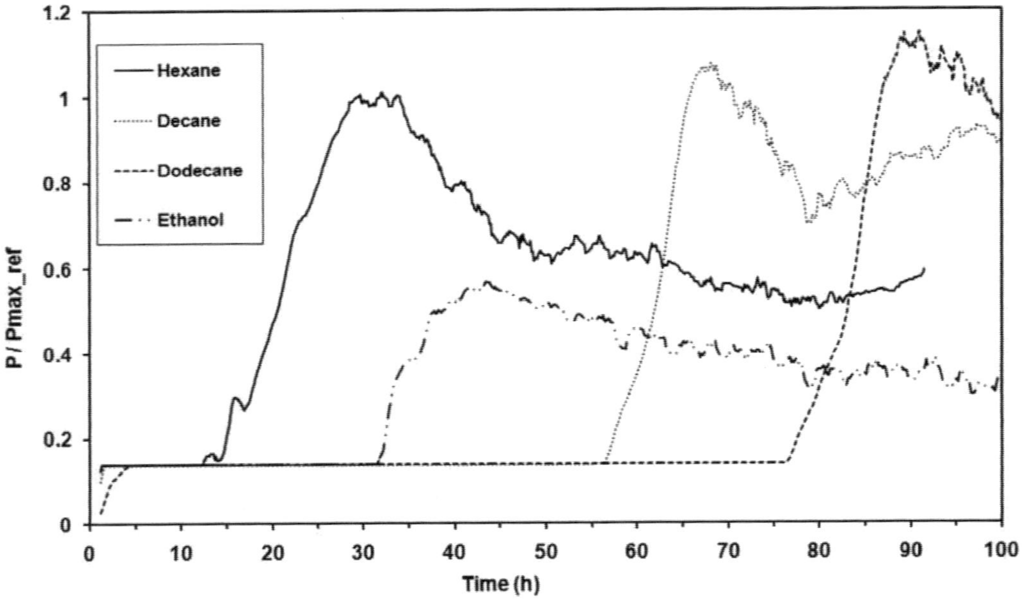

Figure 6. Influence of the substance stored

These curves show that for hydrocarbons, the overpressure is nearly the same in each case studied. The length of the carbon chain has an overall influence on the kinetic of the phenomenon. The pressure raise occurs later for the heavy substances than for the light ones. The case of a low molecular weight alcohol has been modeled too, and the results in this case show that the impact is also on the intensity of the overpressure.

The next parameter studied is the level of the liquid at the beginning of the accident. The scenarios modeled concerned tanks filled with 25, 50 and 90 % of liquid. The gas phase is composed of a mixture of air and vapor, considering the partial pressure of the vapor of the stored substance.

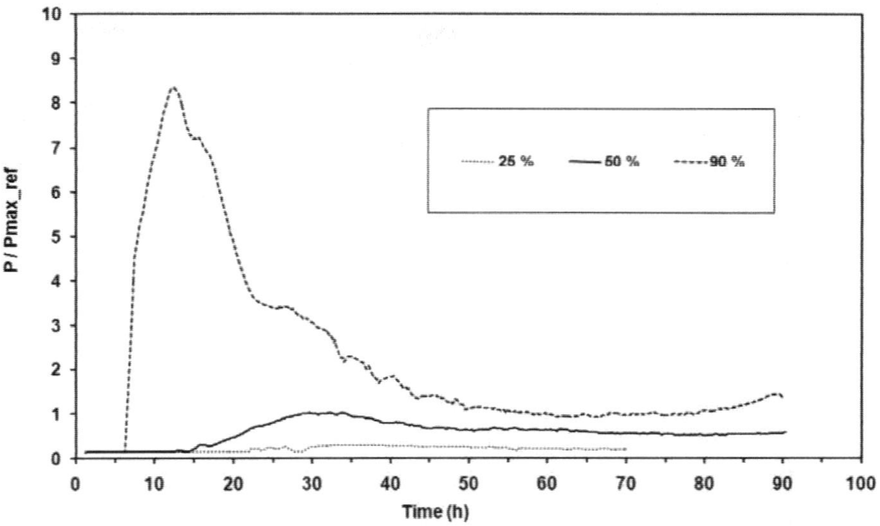

Figure 7. Influence of the tank filling

It seems that an initially full tank leads to a quicker phenomenon, with a really bigger overpressure. This is partially due to the limited amount of gas which is needed to obtain equivalent conditions of pressure in each case. Moreover, the behavior of the system doesn't appear as a linear function of the tank filling. The same study has been done concerning the diameter of the tank:

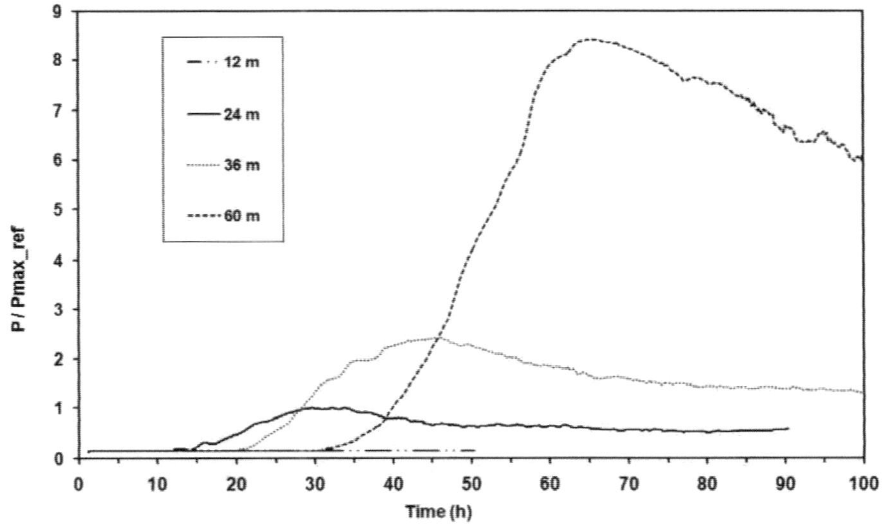

Figure 8. Influence of the tank diameter

Increasing the diameter of the tank has for effect to increase the overpressure obtained. This can be partially explained by the exchange surface between the liquid and gas phase, which increases with bigger diameters. However, as the total volume of liquid increases too, the overpressure is observed later.

The two previous parameters directly impact the area of wetted wall, which seems to be the dominant factor causing the observed peak pressure, due to an efficient heat transfer from the fire. The effect of diameter on time duration seems to also be explained by the ratio wall area / tank volume, which decreases as diameter increases.

The last study done concerned the dimension of the vent. It has overall been done in order to check the behavior of the model.

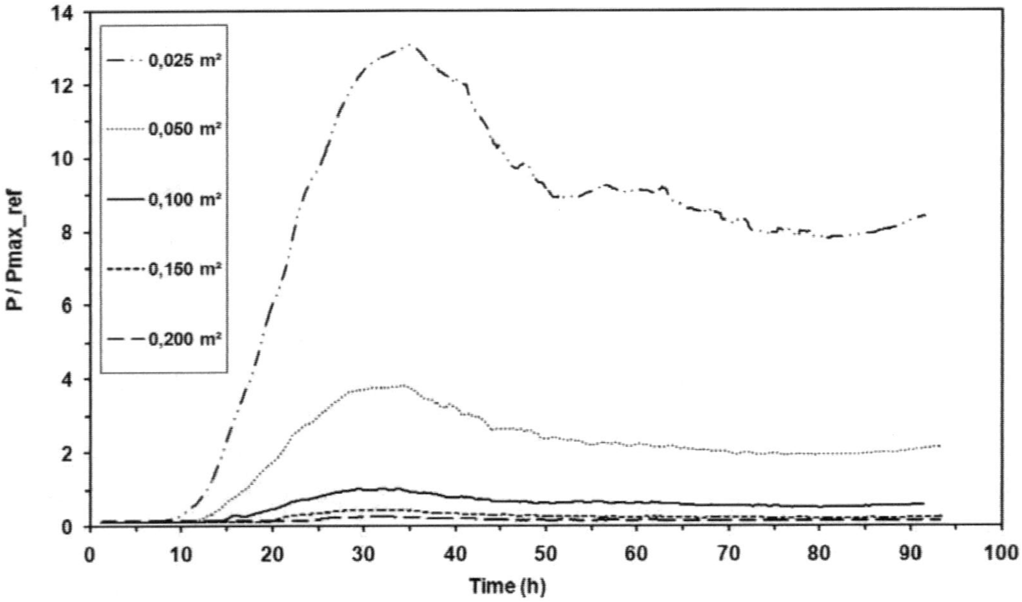

Figure 9. Influence of the vent surface

Results follow the good sense trends: a limitation of the vent size leads to a higher overpressure.

INDUSTRIAL APPLICATION

The results of the previous parametric study have highlighted the predominance of two parameters: the wetted wall area (linked here to the diameter of the tank) and the surface of the vent. In order to establish a general behavior chart, various simulations have been run with different configurations regarding these two input data. The output data will allow to:

- Either in a design phase, calculate the vent size needed to prevent any accident due to pressure increase inside the tank;
- Or for already built tanks, supply input data for the phenomenological study of the potential fireball generated, based on the temperatures mapping in the liquid at the bursting pressure.

For the production of a first general behavior chart, the following conditions were chosen. The fixed parameters are the height of the tank (14 m), the tank wall thickness (10 mm), the filling ratio of the tank (90% of the total height), the substance stored (decane), the maximum radiative flux of the fire (50 kW.m-2) with a progressive loading on the wall. The varying parameters are the vent size and the tank diameter.

A typical result obtained is represented on Figure 10, as a function of time. This curve corresponds to the smallest tank (12 m diameter) with the smallest vent (0,025 m^2) studied.

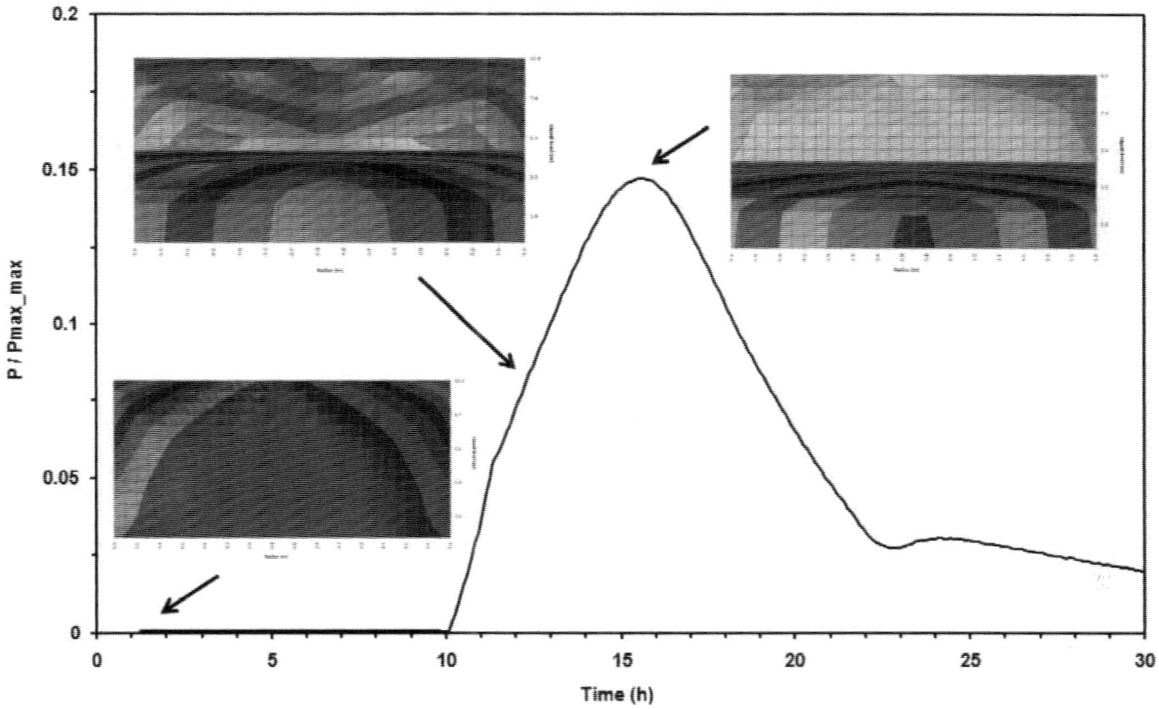

Figure 10. Example of output data

In this Figure, the 2D maps represent the distribution of temperatures in the liquid phase, at different time steps. The "discontinuity" which appears is due to the approach taken to model the fire, with a non uniform loading along the height of the tank. These maps will allow to determine the quantity of liquid which will participate to the potential fire ball at a given failure pressure. Pmax_max is the maximum pressure reached during the parametric study presented in this section, it means for the smallest vent size and maximum tank diameter.

Figure 11 shows the results obtained for the influence of the vent size on the behavior of the pressure, for a 48 m diameter tank. The tank has vents with surfaces varying from 0,025 m^2 to 0,500 m^2.

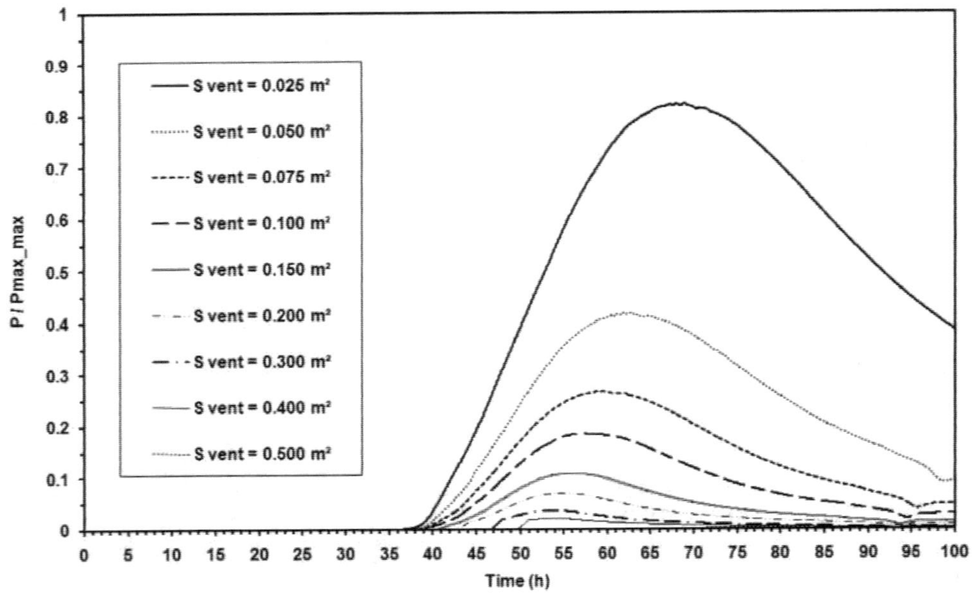

Figure 11. Influence of the vent size for a 48 m diameter tank

The other tanks studied had diameters of 12, 24, 36 and 60 m. By retaining only the maximum pressure obtained during calculations, it is now possible to compile these data in a general behavior chart, as represented in Figure 12.

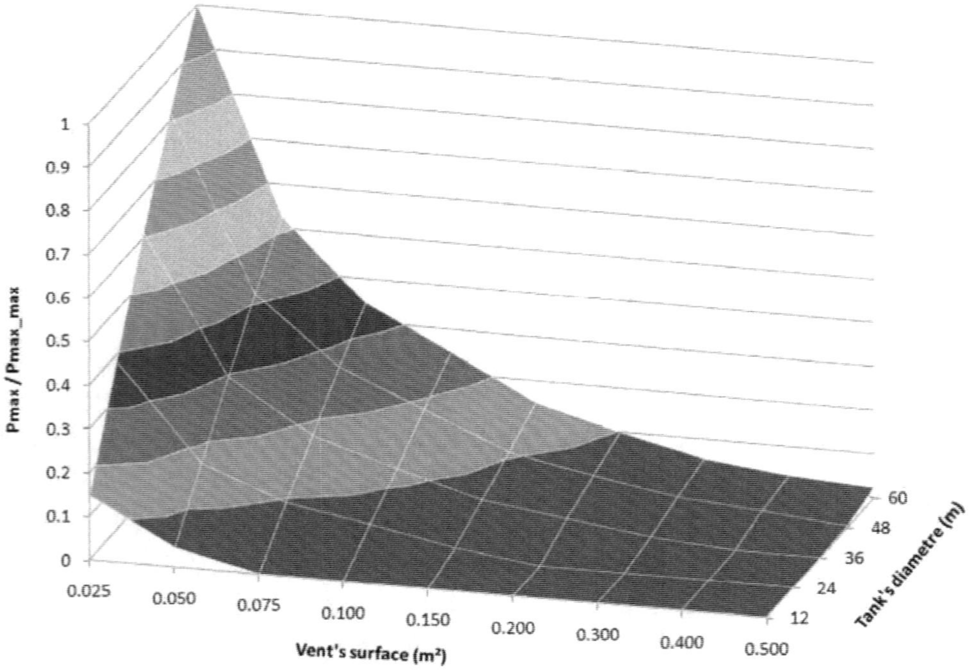

Figure 12. Resulting pressure as a function of tank diameter and vent size

CONCLUSION

The model presented in this paper is a tool developed by INERIS, to determine the surface of vents which should be put in place to compensate the vaporization flow of the content of a fixed roof tank caught in a pool fire. The results obtained with the first parametric studies seem to be consistent with the trends that could be expected. The behavior of the liquid, the heat balance and the kinetic of simulated phenomenon suggest that the computed code works correctly. However, the sensitivity of results to various parameters makes unavoidable the implementation of experimental trials. These parameters have been identified during the development of the model, and the experimental approach will help to adjust them, in order to obtain a model as close as possible to the real behavior. Further to these trials, the next step of this study will be to assess the effects resulting from the rupture of the pressurized tank and from the generation of the fireball.

REFERENCES

1. L. Schuffenecker, J.-N. Jaubert, R. Solimando, « Formalisme et principes de la thermodynamique », Techniques de l'Ingénieur, AF4040, traité Sciences fondamentales.
2. R. Solimando, L. Schuffenecker, J.-N. Jaubert, « Propriétés thermodynamiques du corps pur », Techniques de l'Ingénieur, AF4050, traité Sciences fondamentales.
3. B. Claudel, « Propriétés thermodynamiques des fluides », Techniques de l'Ingénieur, B8020, traité Génie énergétique.
4. American Petroleum Institute (1998), « Venting Atmospheric and Low-Pressure Storage Tanks – Nonrefrigerated and Refrigerated », API Standard 2000, Fifth Edition, Washington D.C., April.
5. American Petroleum Institute (2002), « Manual of Petroleum Measurement Standards » , Chapter 19 « Evaporative Loss Measurement », Section 1 « Evaporative Loss from Fixed-Roof Tanks », API MPMS 19.1, Third Edition, Washington D.C., March.
6. American Petroleum Institute (1993), « Documentation File for API Manual of Petroleum Measurement Standards », Chapter 19 « Evaporative Loss Measurement », Section 1 « Evaporative Loss from Fixed-Roof Tanks », API MPMS 19.1D, First Edition, Washington D.C., March.
7. American Petroleum Institute (1962), « Evaporation Loss From Low-Pressure Tanks », API MPMS 19.1A, API Bulletin 2516, Washington D.C., Marc.

Quantifying the Effectiveness of Explosion Protection Measures

P. E. Moore, *UTC Fire & Security, Mathisen Way, Poyle Road, Colnbrook, Berkshire, SL3 0HB, UK. Phone: +44 1753 689848 Peter.Moore@fs.utc.com*
R. J. Lade, *Kidde Products, Thame Park Road, Thame, Oxfordshire,OX93RT, UK. Phone: +44 1844 265117 Rob.Lade@Kiddeuk.co.uk*

1. INTRODUCTION

In industrial processing of flammable materials including combustible dusts there is an attendant risk of ignitions leading to explosion consequences. A key process plant design objective is to minimise the risk and consequence arising from any ignition by the incorporation of efficacious explosion prevention and explosion protection measures.

There exists a substantive literature that details the effectiveness of explosion protection measures and robust guidance that leads to viable and secure explosion safety at the plant component level. Much of this detail is engrossed in our NFPA Codes [1-4] that establish a best practice framework for industry.

In practice design engineers are challenged by the scale of this task – particularly when considering the complexity of many of our industrial processes. The challenge is compounded because the design demand is not always intuitive or apparent. The authors have advocated the use of a risk modelling tool which sets out to quantify and interrogate the standing residual risk that the elected safety measures would fail to mitigate an explosion consequence. This approach can assist practitioners in ascribing efficacious explosion protection.

2. THE CALCULATION TOOL

The calculation methodology and its strict mathematical interpretation has been previously described [5-7] and is summarised herein. It provides for a representation of any process by a set of vertices (the processing vessels) and connections (all of the ducts and communition equipment) between these vertices.

2.1 Directed Graph Representation

Process plants under study are represented by a connected, directed graph [8]. In the representational architecture, each vessel in the system is defined as a vertex. In the event of an ignition, the edges between the vertices represent possible paths of flame propagation (e.g. duct-work between connected vessels). Between any pair of adjacent vertices, there are two directed edges in opposite directions. Each edge is associated with a weighting which represents the directional probability of flame propagating along the connection. Figure 1 shows a schematic of a typical spray drying process which Date et al [7] described to clarify the mathematical basis of this deterministic method. Here a wet dairy product is spray dried, and then passes through two fluid bed driers that further reduce the moisture content of the final product. Dust content in the drying air is separated by a ganged pair of cyclones, and returned via a fines return line to the spray drier.

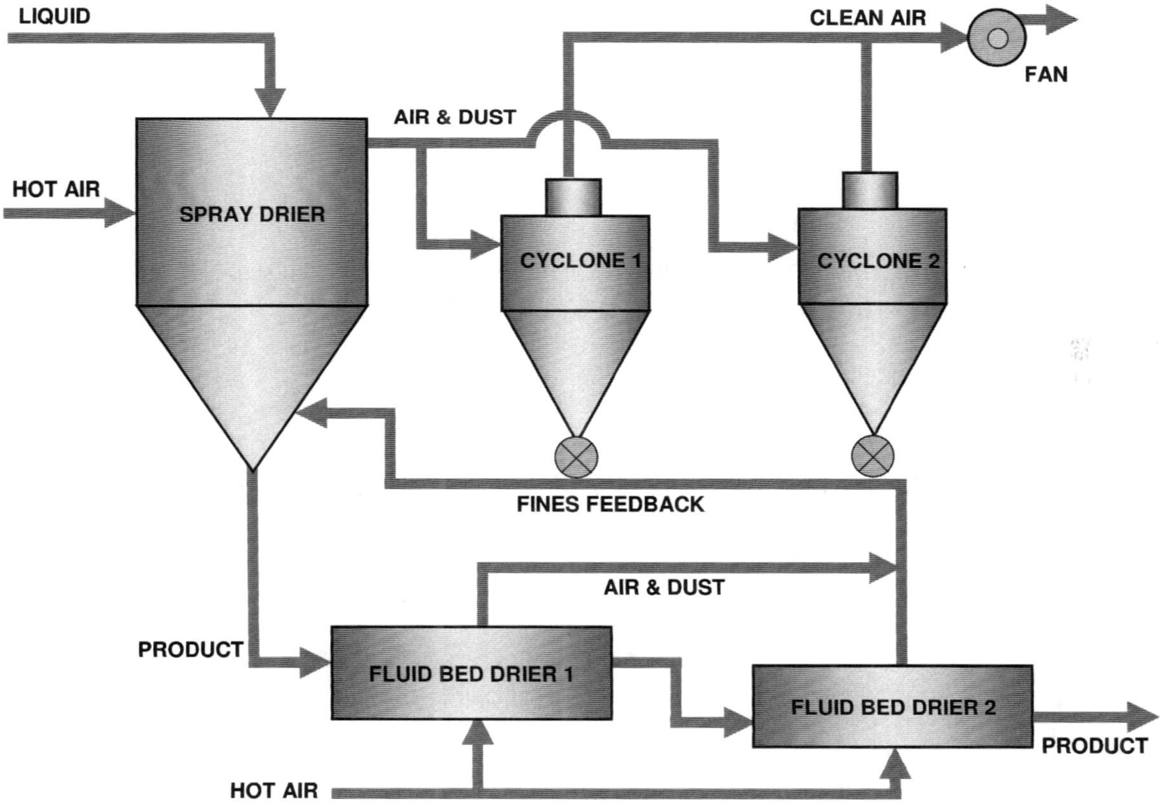

Figure 1: Schematic representation of an example spray drying process. The arrows represent the flow directions through the process plant.

Figure 2 shows the corresponding directed graph representation for this process*. Vertices are just abstract representations of the plant vessels, and each vertex has multiple edges.

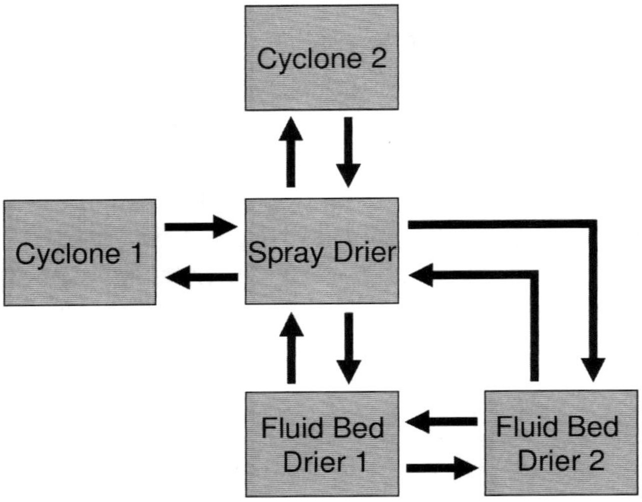

Figure 2: The directed graph representation of the spray drying example.

2.2 Description of the Calculation Methodology

- The explosion challenge relies on the measured explosibility parameters† of P_{max} and K_{max}, which are considered as representative worse case for the application in question.

- An unmitigated explosion (failure) is defined as any occurrence at any vertex where the reduced explosion pressure, P_{red}, of a suppressed or vented explosion is greater than the pressure shock resistance, P_s, of the plant item – i.e the design inequality $P_{red} \leq P_s$ is **NOT** met.

- Explosion propagation down a connection is considered unmitigated in any occurrence where the elapsed time to establish an explosion isolation barrier, t_b,

---

* In this representation it is assumed that the probability of deflagration propagation through the rotary gate valves between the fines return line and the cyclones C1 and C2 is zero.

† Industry practice [9,10] is to quantify material explosibility by a measurement of the maximum explosion pressure, P_{max}, and the explosibility rate constant $K_{max}=(dP/dt)_{max}.V^{2/3}$ of a turbulent explosion in a closed volume, V.

is greater than the time to transition flame to the barrier location, t_f – i.e. the design inequality $t_b \leq t_f$ is **NOT** met.

- Given an ignition event at a vertex, an unmitigated explosion is assumed to occur when any one component[‡] of the explosion protection system fails, be it an explosion vent device, explosion detector, explosion suppressor or control panel. The method uses each component's reported MTBF[§] (mission critical failures only) as a proxy for their risk of failure.

- The model treats all failures equally and ascribes no consequence specific to the failure scenario. In reality many of the envisioned failures will only represent inconvenience (such as some minor plant distortion) but some would be catastrophic.

- In the cases where there are multiple flame paths between adjacent vessels, these paths are considered independently and the total probability of flame transfer determined by summing over all paths.

2.3 The Calculation Basis (Overview)

Each vessel or plant item i (vertex i) within the process plant, together with its associated explosion protection system is characterised by a set of parameters.

$Q_E(i)$ is the ignition probability in vertex i. For a given process plant and over a given unit of time we assume that $\sum_i Q_E(i) = 1$, i.e. that there will be one ignition occurrence somewhere in the process plant against which we seek to determine the residual risk that the resulting explosion will not be mitigated at one (or more) of the vertices.

The risk of failure of any vertex i due to ignition in vertex j, is denoted $R_{i,j}$ and can be computed as the sum of the risk of hardware failure, $Q_h(i)$, and the risk of inadequate explosion protection (the second term in Equation 1):

$$R_{i,j} = Q_h(i) + (1 - Q_h(i)) \times Q_{vessel}(i,j) \qquad \text{Equation 1}$$

‡ Designs with any hardware redundancy require a failure of both components.

§ MTBF – Mean Time Between Failure

where $Q_{vessel}(i,j)$ represents the proximity of P_{red} to P_s in the system design and accounts for any intentional design safety factors in our computation of residual risk; and $Q_h(i)$ can be calculated from Equation 2.

$$Q_h(i) = \alpha(i) + (1-\alpha(i)\beta(i) + (1-\alpha(i))(1-\beta(i))\gamma(i) \qquad \text{Equation 2}$$

The terms in Equation 2 for $Q_h(i)$ represent the hardware failure exposures of vent panels, α, explosion detectors, β, and explosion suppressors, γ as installed as part of the explosion protection measures. $Q_h(i)$ as a whole represents the probability that an unmitigated explosion occurs in vessel i due to hardware failure.

The risk of failure of any vertex due to an ignition in vertex i may be denoted by δ_i and can be computed as:

$$\delta_i = Q_E(i)\left(R_{i,i} + (1-R_{i,i}) \sum_{j \in \Phi_i} Q^s(i,j) R_{j,i} \right) \qquad \text{Equation 3}$$

where Φ_i denotes the set of vertices adjacent to vertex i and $Q^s(i,j)$ represents the total flame propagation probability from vessel i to j and will be dependent on the geometric configuration and the explosion hazard, together with the reliability of any explosion isolation hardware.

Each $R_{j,i}$ is computed using Equation 1. Note that the first term in Equation 3 represents an event where an ignition in vertex i causes an unmitigated explosion in the vertex i and the second term with the summation over j represents an event where there is no unmitigated explosion in vertex i given an ignition in the same vertex, however, the flame propagates to a connected vertex j causing an unmitigated explosion in vertex j.

Instead of computing the "per-ignition" risk (due to ignition in vertex i), δ_i, one may choose to compute the "per-vertex" risk, i.e. total risk of failure in each vertex due to ignition in the same vertex or in any of the connecting vertices. Denoting this risk with ς_i, we arrive at Equation 4

$$\varsigma_i = Q_E(i)R_{i,i} + \sum_{i \in \Phi_j} Q_E(j)(1 - R_{j,j})Q^s(j,i)R_{i,j}$$ Equation 4

The overall residual risk R can be computed via Equation 5 where we have chosen to sum over ς_i. A summation over δ_i with appropriate inclusion of the probability of failure of the suppression system control panel would yield the same results as Equation 5.

$$R = \sum_j \left\{ \pi(j) + (1 - \pi(j)) \sum_{i \in \Psi_j} \varsigma_i \right\}$$ Equation 5

where π is the probability of failure of the suppression system control panel. The summation is over all control panel zones and $\Psi_1, \Psi_2, \ldots$ represent the MTBF for each control zone.

3.0 QUANTIFYING EXPLOSION PROTECTION EFFECTIVENESS

To assist in our appreciation of the value of such analyses, let us consider a plant segment of just two connected vessels as shown in Figure 3 – each with an explosion vent sized in accordance with the prevailing NFPA codes.

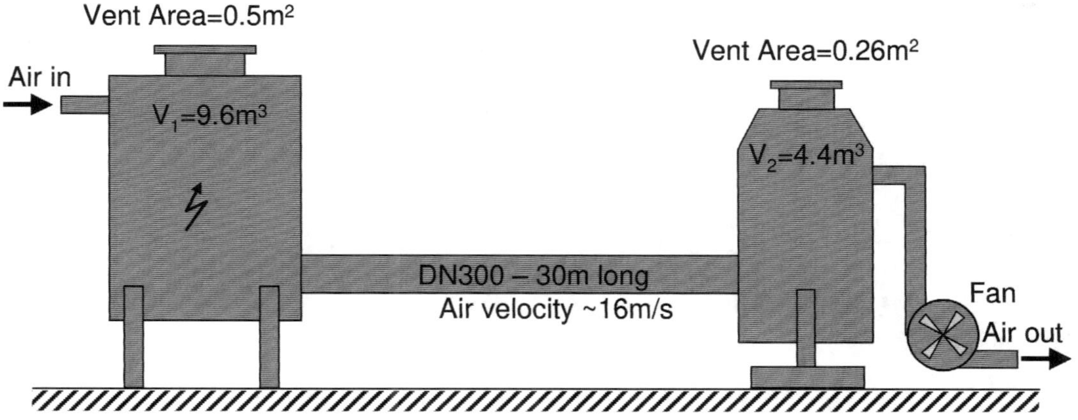

Figure 3: Schematic representation of a 9.6m^3 (V_1) connected via a 30m long DN300 duct to a secondary vessel V_2 (4.4m^3). Both V_1 and V_2 have a pressure shock resistance P_s~0.6bar and vent areas calculated based on a material explosibility of K_{max}=160 bar m/s.

We know that for such a configuration with an ignition in V_1, there is a high risk of flame propagation from V_1 to V_2 [11,12], and that such a connected vessel explosion will result in a very intense explosion in V_2. Strict application of the current explosion venting guidance, however, would prescribe a $0.26m^2$ vent panel on V_2 because it would be evaluated independently as a stand alone vessel. Application of the described tool with the datum set of input values to this scenario delivers up a calculated (theoretical) residual risk of the protection (in this case the two explosion vent panels) failing to mitigate, see Table 1.

Table 1: Calculated residual risks for vertices V_1 and V_2 in Figure 3 assuming an equal ignition probability in both vertices.

Plant Component	Residual Risk ς_i
V_1	8.7×10^{-2}
V_2	1.1×10^{-1}
Overall Residual Risk, R	1.97×10^{-1}

The overall residual risk that the explosion protection measures fail to mitigate is the arithmetic sum of the risks at each vertex of the plant segment. In this case the analysis teaches that there is around a 1 in 5 chance that any ignition will result in an explosion that is not fully mitigated, which is entirely reasonable because of the high probability of a flame propagation between V_1 and V_2 leading to enhanced connected vessel explosions.

The inclusion of an explosion isolation measure between these two vessels would be expected to significantly reduce this residual risk. Figure 4 shows the application of a triggered extinguishing barrier installed along the 30m interconnection.

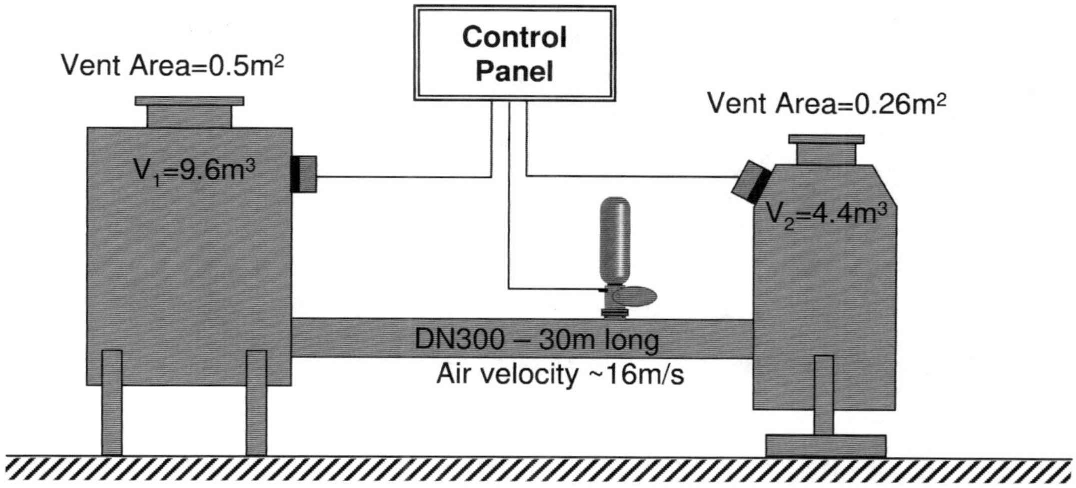

Figure 4: Schematic representation of a 9.6m^3 connected via a 30m long DN300 duct to a secondary vessel V_2 (4.4m^3) whereby pressure detectors are employed to trigger a suppressant barrier on the connecting duct work.

Table 2 shows that this additional explosion protection expedient reduces the residual risk of an unmitigated explosion by almost two orders of magnitude.

Table 2: Calculated residual risks for vertices V_1 and V_2 in Figure 4 assuming an equal ignition probability in both vertices.

Plant Component	**Residual Risk** ς_i
V_1	1.4×10^{-3}
V_2	1.2×10^{-3}
Overall Residual Risk, R	2.6×10^{-3}

Date et al. [7] analysis of a more complex application – the spray drier shown in Figure 1 with the envisioned explosion protection measures and logical zones of protection shown in Figure 5 - deconvolutes risk from a perspective of per ignition in each vertex, δ_i, and per vertex ς_i – see Table 3.

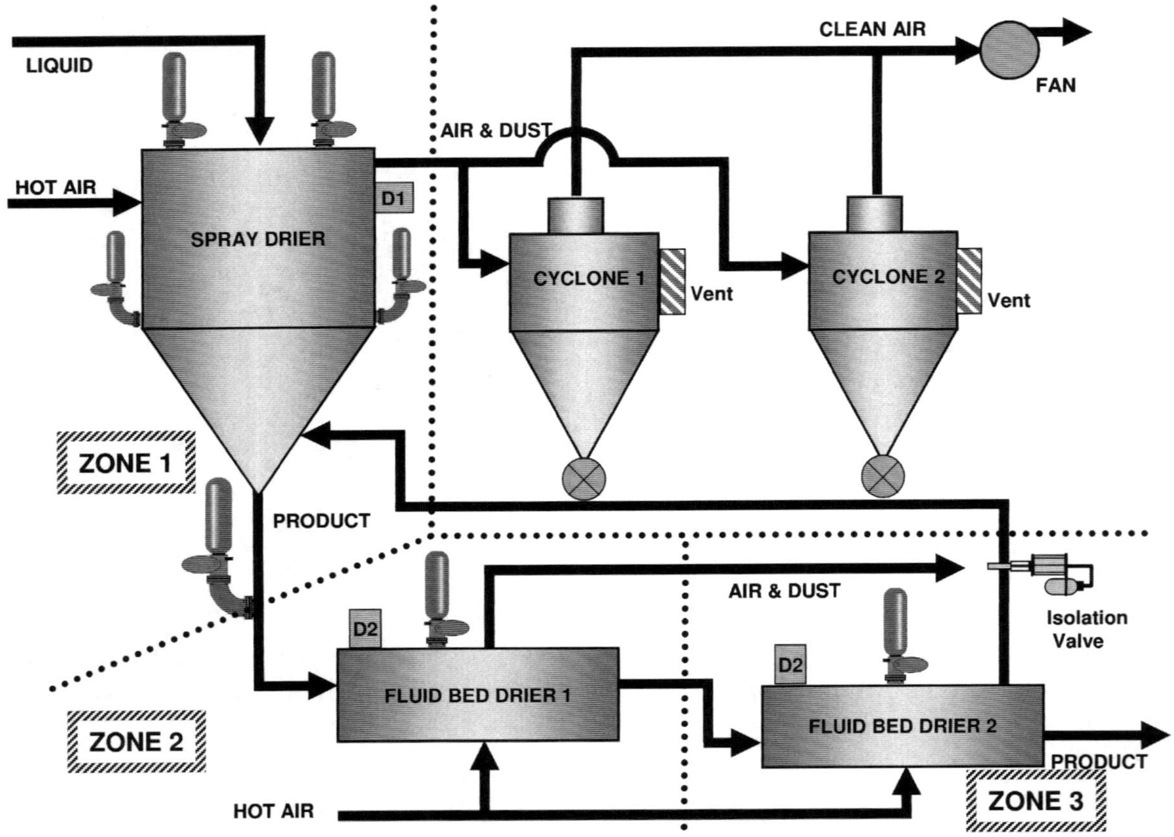

Figure 5: Schematic of a simple spray drier process with envisioned explosion protection measures.

Table 3: Calculated residual risks for all for vertices shown in Figure 5 on a per ignition (δ_i) and per vertex basis (ς_i)

Vertex	δ_i (ignition basis)	ς_i (vertex basis)
Spray Drier	3.41×10^{-3}	1.11×10^{-4}
Cyclone 1	6.49×10^{-3}	7.95×10^{-3}
Cyclone 2	6.26×10^{-3}	7.77×10^{-3}
Fluid Bed Drier 1	3.14×10^{-3}	4.64×10^{-4}
Fluid Bed Drier 2	3.14×10^{-3}	4.64×10^{-4}
Overall Residual Risk, R		1.68×10^{-2}

Such analyses are specific to the detail of the process and the pertinence of the input parameters. They can deliver up a systematic de-convolution of the overall residual risk by identifying relative risk contributions, and can assist in driving down such risk by a relentless pursuit of the higher risk elements of the application.

Lade et al [6] described a milling process with envisioned explosion protection measures and de-convolutes the residual risk of failure to mitigate using this method, see Figure 6 and Table 4. For this example we see a well protected plant segment where each of the vertices has an equivalent SIL2 (safety integrity level) or higher rating, but the overall process segment itself is only at SIL1.

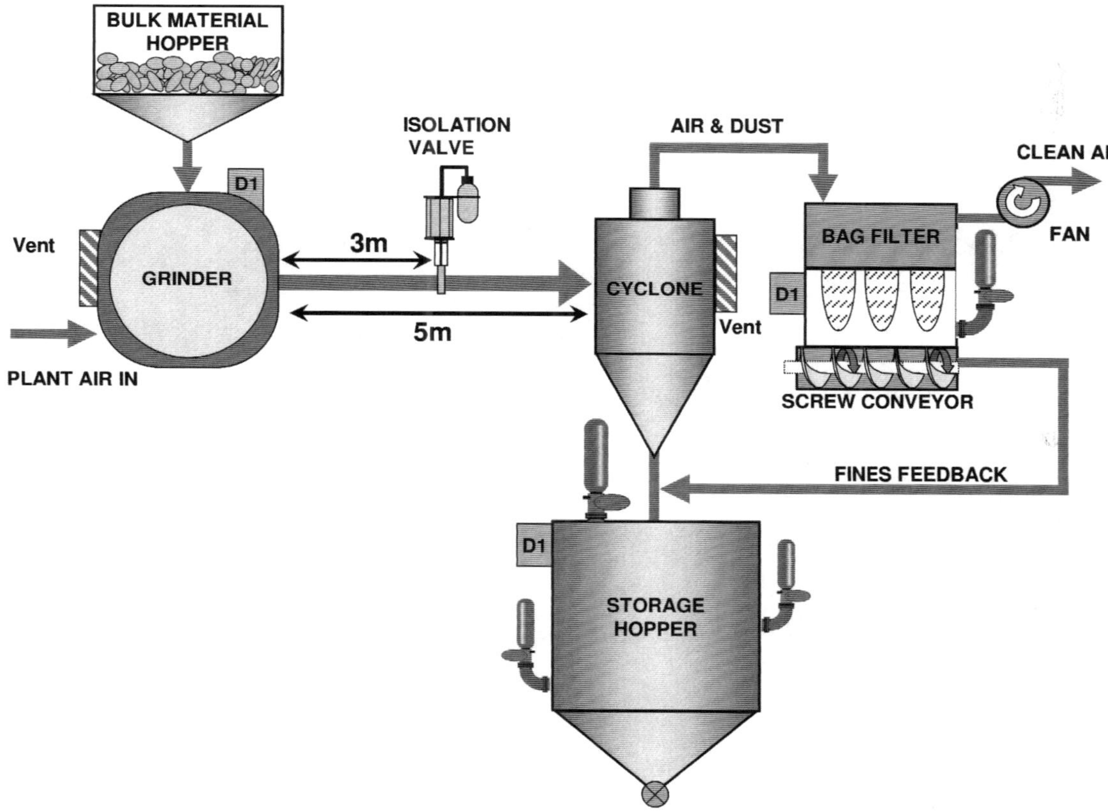

Figure 6: Schematic representation of a milling and collection process [6] with envisioned explosion protection measures.

Table 4: Calculated residual risks for the application shown in Figure 6 on a per vertex basis (ς_i) together with the overall risk

Vertex	ς_i (vertex basis)
Grinder	2.79×10^{-3}
Cyclone	4.95×10^{-3}
Bag Filter	7.10×10^{-3}
Storage Hopper	5.43×10^{-5}
Overall Residual Risk, R	1.50×10^{-2}

4.0 CHALLENGING THE IMPLICIT ASSUMPTIONS

If reliance is to be placed on a risk model of this type, then the validity of its assumptions must be understood. The risk model accounts for:

- Hardware failure
- Ineffective protection of vertices – where the design inequality $P_{red} \leq P_s$ is compromised
- Ineffective protection of connections – where the design inequality $t_b \leq t_f$ is compromised

It derives a residual risk that the protection fails to mitigate based on the contributions of the risks of the above, but does not represent any consequence of said failures.

It is clearly the case that a calculation of residual risk is mathematically tractable for any reasonable set of assumptions, and that the value of the output is therefore dependent on the rationale of the assumptions. It is important, therefore, to review the assumptions, consider their validity and their sensitivity to the output projections. The basis of this approach to ascribing efficacious explosion protection is to assume that there will be an ignition occurrence, somewhere in the process segment under consideration, and to use the derived probability that the resultant explosion will not be mitigated as a proxy for ascribing residual risk.

Clearly for any deterministic measure of the exposure it is necessary to condition a derived residual risk by the expected frequency of occurrence of ignitions. This is somewhat intangible. It will be a function of the type of process, the scale and complexity of the process, and the rigour of control and maintenance practices. Ascribing the residual risk that an ignition will lead to an unmitigated explosion consequence is thus just a first step in the determinant. We examine below the assumptions within the calculation method.

4.1 Ignition Location Assumption

The described model considers only ignition scenarios at vertices. Any ignition in an interconnection is expected to become an ignition at one or more vertices, and the outcome (risk of failure to mitigate) is thus the consequence of said ignition(s). This represents a reasonable basis to proceed provided that an ignition in an interconnection will not result in pressure or flame damage to the interconnection prior to its propagation to a vertex. This assumption would breakdown for long interconnections (where flame acceleration could transition to a detonation).

4.2 Ignition Probability Assumption

The described model relies on the historical evidence of ignition occurrences to ascribe the relative probability of ignition based on the vertex duty cycle. Clearly processes vulnerable to mechanical spark, frictional heating and/or glowing ember ignitions have a higher incidence of ignition relative to processes such as dust separation and pneumatic transport. One such approximation extracted from the data reported in [13,14] is shown in Table 5.

Table 5: Normalised ignition probabilities for different generic plant processes.

Process Category	Example Plant	Normalised Ignition Probability
Mechanical working, high speed	Micronisers, Pin mills, Grinders, Hammer mills	31%
Mechanical working, low speed	Communition equipment, Elevators, Fans	25%
Forced heating	Spray, bed, ring dryers, T>100°C	25%
Dust separation	Filters, Sieves, Separators, Classifiers	9%
Pneumatic transport	Cyclones, Transport ducts	7%
Bulk storage	Silos, Bins, Hoppers	3%

Note that this simple treatment makes no allowance for the scale of the respective plant components and also presumes that any ignition is equally probable at all locations within the said component. Risk models would benefit from a more rigorous quantification of ignition probabilities.

4.3 Parametric Uncertainty Assumption

The nominated values of each of the key parameters P_{max}, K_{max}, P_{red}, P_s, t_b and t_f are assumed to represent the 2σ limit value from a normal distribution (i.e. 95% confidence that said elected values are not understated), and the default value for the standard deviation, σ, is set as 10% of the parametric mean.

The examination of representative experimental data sets - in this case the measured explosibility of a turbulent starch dust in a 6.2m^3 vessel, and the suppression of this explosion using a proprietary explosion suppression system** demonstrates the reasonableness of this assumption, see Table 6.

Table 6: Experimental data for unsuppressed and suppressed high turbulence maize starch explosions in a 6.2m^3 vessel.

	P_{max}	K_{max}	P_{red}
Type of Test	Unsuppressed	Unsuppressed	Suppressed
Number of tests / n	6	6	4
Mean value / x	8.66 (bar)	302 (bar m s^{-1})	0.49 (bar)
Standard deviation / σ	0.21 (bar)	17.7 (bar m s^{-1})	0.07 (bar)
x/σ	2.4%	5.9%	14.8%

It is often the case that the uncertainty in the quantification of pressure shock resistance, P_S, for plant components is greater than this default value, and in such cases practitioners would be advised to elect a more conservative criterion.

4.4 Connectivity Assumption

The risk is computed over all flame paths between each of the adjacent (directly connected) vertices. For each connection there are three alternative outcomes:

- Flame propagates from the ignition location (vertex i) to the connected vertex (vertex j) and causes a sympathetic explosion in the connected vertex that is more intense than would have been the outcome of a point ignition in vertex j. This consequence is referred to as "flame jet ignition" and arises because the connected vertex experiences both pre-compression from the upstream explosion and a jet of flame blasted into the vertex as an ignition scenario [11,12]. The consequence is an intensified explosion that is

** Two 16kg explosion suppressors with a 75mm outlet charged with KIDDEx™ suppressant with an actuation pressure of 0.1bar(g).

substantially more challenging for any installed explosion protection on this vertex; or

- Flame propagates from the ignition location (vertex i) to the connected vertex (vertex j) with little blast effect and gives rise to a sympathetic explosion in the connected vessel that is no more intense than would have been the outcome of a point ignition in that connected vertex; or

- Flame propagation collapses in the connection and there is no sympathetic ignition in vertex j.

The purpose of an explosion isolation measure such as a triggered gate valve or an extinguishing barrier is to minimise the risk of flame propagation, and thus the occurrence of a sympathetic explosion. In the case where such isolation fails because the criteria $t_b \leq t_f$ is not met, the action of the barrier will nevertheless substantially reduce the risk of a flame jet ignition induced enhanced explosion because it reduces the blast impact, and in the case of the extinguishing barrier may provide some inerting of vertex j.

Risk models are reliant on the corpus of experimental data, and validated computation fluid dynamic (CFD) modelling tools like FLACS [15] (and the dust variant DESC) to assist in ascribing the flame connectivity expectation. They would clearly benefit from an increased corpus of test data and/or improved CFD treatments. Provided that connectivity assumptions are reasonable, however, risk models can deliver systematic projections of the residual risk map.

4.5 First Order Assumption

The described calculation method considers that for any ignition in a vertex i only the probabilities of an unmitigated explosion at this vertex and in all adjacent connected vertices j, contribute to the risk of failure. Thus the only flame paths considered are those between directly connected vertices.

The calculation method is fully tractable for secondary or tertiary order assumptions. It can be shown that in practice this extension has a minimal impact on the derived residual risks.

5.0 BENEFITS TO OVERALL PROCESS SAFETY INTEGRITY LEVEL (SIL)

Quantifying the efficacy of explosion protection measures and their impact on the residual risk that pertains is complex. The use of a systematic methodology has been described, and the assumptions implicit in such a method explored. Whilst there is a necessity to accept certain assumptions, the method provides a systematic means to permit the comparison between explosion protection options, and importantly, to assist the design engineer to focus on those aspects of any protection system that make the greatest contribution to residual risk. The benefits of explosion protection options are not always intuitive, and by the adoption of a systematic means to ascribe risk better choices are more probable.

It could be argued that the method is restrictive simply because of the disconnect between risk and consequence – clearly the primary aim is not so much to reduce risk of failure but to reduce risk of high consequence failure. To that end it is evident that a quantification of the overall risk, and most importantly the major contributors to this risk, is a valuable step. Moreover, as explained above, the risk determinate does not address the absolute frequency of expected ignition occurrences, and its interpretation must be conditioned by this consideration.

The described method herein permits a means to quantify risk of vertex failure (irrespective of where ignition is envisioned), and can be used to assist in ascribing efficacious explosion protection. In the spray drier example the greatest consequence (both risk to personnel and cost of business recovery) would be substantive damage or destruction of the very large spray drying tower. Although, relative to the risk of failures at other vertices, this is already a low exposure (in this case 1.1×10^{-4}), it would be prudent to consider actions that could further reduce this contributing risk. In the example, the spray drier is fitted with an explosion suppression system and the key contributors to this risk element will be the hardware's intrinsic reliability and the safety factor between the expected worse case suppressed explosion pressure, P_{red}, and the tower's internal pressure shock resistance, P_s. The former can be addressed by the selection of proprietary hardware that has higher intrinsic reliability and/or by the inclusion of some redundancy. The latter can be addressed by changing the suppression system design to reduce P_{red}, or by specifying/strengthening the tower construction for a higher P_s. Table 7 below shows the theoretical impact of such change options that would form part of the users' final consideration and election of operational practice for this application.

Table 7: Impact of residual risk of the spray drier shown in Figure 6 with re-specification of hardware reliability and re-specification of reduced explosion pressure, P_{red}.

	ς_i
Spray Drier ς_i datum – see Table 3	1.11×10^{-4}
Revision A: Re-specification of explosion suppression system with higher reliability components such that hardware MTBF is increased by a factor of 2 on this vertex.	5.55×10^{-5}
Revision B: Re-specification of explosion suppression system such that the expected worse case P_{red} = 50% P_s	7.89×10^{-5}
Impact of BOTH Revision A and B.	3.45×10^{-5}

Interrogating the risk of failure at each vertex assists in opining on consequence. The evolutionary next step for industrial explosion protection would be to migrate today's capability towards an "expert system" design tool that can maximise safety provisions by taking account of both the risks and consequences and deliver up robust cost/benefit projections. The authors contend that this evolution is now within sight.

6.0 ACKNOWLEDGEMENTS

The authors would like to thank Professor Gautam Mitra and Dr Paresh Date from the *Center for the Analysis of Risk and Optimisation Modelling Applications*, Brunel University, UK, for their work in devising the mathematical methodology employed for the computation of residual risk.

7.0 REFERENCES

1. NFPA 61: (2008) Standard for the Prevention of Fires and Dust Explosions in Agricultural and Food Processing Facilities, National Fire Protection Association, Quincy, MA.

2. NFPA 654: (2006) Standard for the Prevention of Fire and Dust Explosions from the Manufacturing, Processing, and Handling of Combustible Particulate Solids, National Fire Protection Association, Quincy, MA.

3. NFPA 68: (2007) Standard on Explosion Protection by Deflagration Venting, National Fire Protection Association, Quincy, MA.

4. NFPA 69: (2006) Standard on Explosion Prevention Systems, National Fire Protection Association, Quincy, MA.

5. Ganguly, T., Date, P., Mitra, G., Lade, R. J. and Moore, P. E., (2007), "A method for computing the residual risk of safety system failure" in `Proceedings of 12th International Symposium on Loss Prevention and Safety Promotion in the Process Industries'

6. Lade, R. J. and Moore, P. E., (2008), "A methodology to guide industrial explosion safety system design", Hazard XX: Process Safety and Environment Protection' IChemE No 154.

7. Date, P., Lade, R. J., Moore, P. E., and Mitra, G., (2009), "Modelling the Residual Risk of Safety System Failure", submitted to Journal of Loss Prevention.

8. West, D. (2001), *Introduction to Graph Theory*, Prentice Hall.

9. ISO 6184: Parts 1-3 (1985) Explosion Protection Systems – Determination of Explosion Indices of Combustible Mixtures in Air.

10. ASTM E1226, Standard Test Method for pressure and rate of pressure rise for combustible dusts, ASTM International, Conchehocken, PA.

11. Holbrow, P., Andrews, S., Lunn, G. A., 1996, "Dust Explosions in Interconnected Vented Vessels", J Loss Prevention Process Industries, 9(1): 91-103.

12. Lunn, G. A., Holbrow, P., Andrews, S., Gummer, J., 1996, "Dust Explosions in Totally Enclosed Interconnected Vessel Systems", J Loss Prevention Process Industries, 9(1): 45-58.

13. Berufsgenossenschaftliches Institut für Arbeitssicherheit – Report NR. (1982) 4/82 Dokumentation Staubexplosionen, Analyse und Einzelfalldarstellung, ISSN 0173-0487.

14. Berufsgenossenschaftliches Institut für Arbeitssicherheit – Report NR. (1997) 11/97 Dokumentation Staubexplosionen, Analyse und Einzelfalldarstellung, ISSN 0173-0387 (ISBN 3-88383-463-7).

15. FLACS: (FLame ACceleration Simulator), (2005) Gexcon AS, Bergen, Norway.

83e The Failure to Communicate: Risk Assessment Challenges Outside the Chemical Process Industries

Catherine Kollhoff
C. S. Howat & Associates
Chicago, IL 60622
Katie.Kollhoff@gmail.com

Colin S. Howat
University of Kansas
Lawrence, KS 66045-7609
cshowat@ku.edu

Abstract:

The chemical process industries (CPI) have developed sophisticated methods to identify hazards, identify their avenues for escape, develop methods to block their escape and assess the associated risk. These methods were spawned by notable world-scale excursions resulting in substantial loss of life, capital and business. These methods assume a level of personnel sophistication. In fact, they have spawned an educational system that prepares analysts to look beyond intended performance rooted in technical fundamentals.

World-scale methods applied to neighborhood industries and economics result in a disconnect between the assumed and actual outlook, education and sophistication. This disconnect challenges the direct implementation of the world-scale methods. We only have look at Imperial Sugar to recognize that the methods of the CPI have not gone beyond that industry. Hazards go unrecognized, near-hits are not investigated, risk is not evaluated and world-scale events are the result.

Bringing CPI methods to a perspective and engineering-fundamentals challenged culture results in failure to identify properly hazards and mitigate risks. The initiative spawned by CSB to force OSHA to regulate an industry far outside the traditional PSM/PHA focus will be problematic until methods are tailored to those industries to educate them in fundamentals foundation, hazard perspective and excursion prediction.

This paper presents our experiences over the last five years to evaluate risk in an industry far outside the traditional CPI. Specific examples in communication breakdown will be presented to form the foundation for proposed evolutionary procedure, discussed in the paper, which not only educates the plant personnel but also identifies hazards, evaluates consequence and mitigates risk.

Are our facilities safe?... And now?... And now?... Safety assurance in large installations with constant changes

Bruno, Dante – Pan American Energy – Alem 1180, Buenos Aires, Argentina.
Alpert, Melina; Di Campli, Yanina; Dominguez, Juan Carlos – TECNA S.A. – Encarnación Ezcurra 365, Buenos Aires, Argentina.

ABSTRACT

Constant changes in production conditions, demand constant changes in installations, and although changes may (or not) be adequately registered, one has to wonder... with all those changes we made, are our facilities still safe?

This is the case of an integrated facility having six compressor plants, three gas processing plants, three flares, more than 85 km of interconnecting gas pipeline, over more than 60 km^2.

The analysis involved qualitative risk analysis sessions to determine scenarios. Following, those scenarios were analyzed, with both steady state and dynamic models and considering two flow situations, actual and future, so as to be able to identify how safeguards would behave. These analyses were primary aiming at avoiding overpressure and avoiding flaring rates beyond capacity.
Finally, conclusions and recommendations were made.

Summarizing, this paper reviews how hazards and scenarios were identified in this very complex installation, and how different existing safeguards were analyzed to decide if they were suitable or not to achieve integrity assurance.

INTRODUCTION

Pan American Energy (PAE) is the second largest producer of hydrocarbons in Argentina. It was created by BP & Bridas in 1997 to develop the energy business in the South American region. The "Golfo San Jorge" basin, the area of interest for the present study, produces more than 100000 bbl/d (15900 m3/d) of crude oil and more than 247 MMSCFD (7 MMSMCD) of gas.

Tecna is a global engineering and constructions company, with more than 700 employees and more than 30 years experience in the energy business. Tecna has contributed with PAE in this study.

SYSTEM DESCRIPTION

Basic gas system is shown below.

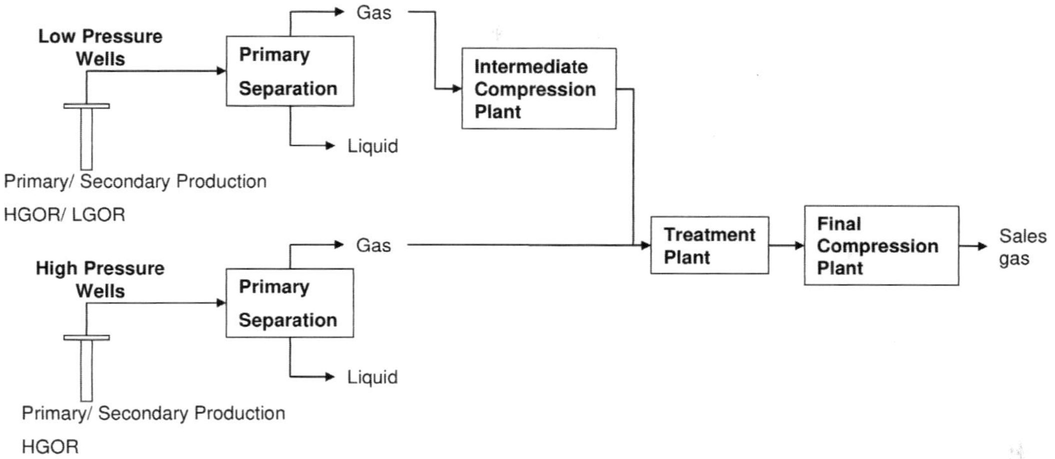

Figure 1. Basic Gas System.

Detailed gas system is shown below. It is a complex pipe network and multiple interconnections between installations are possible. One event, in one of the Gas Treatment Plants, may affect not only the compressor plant feeding facility but also all other upstream and downstream facilities connected by the same pipe network.

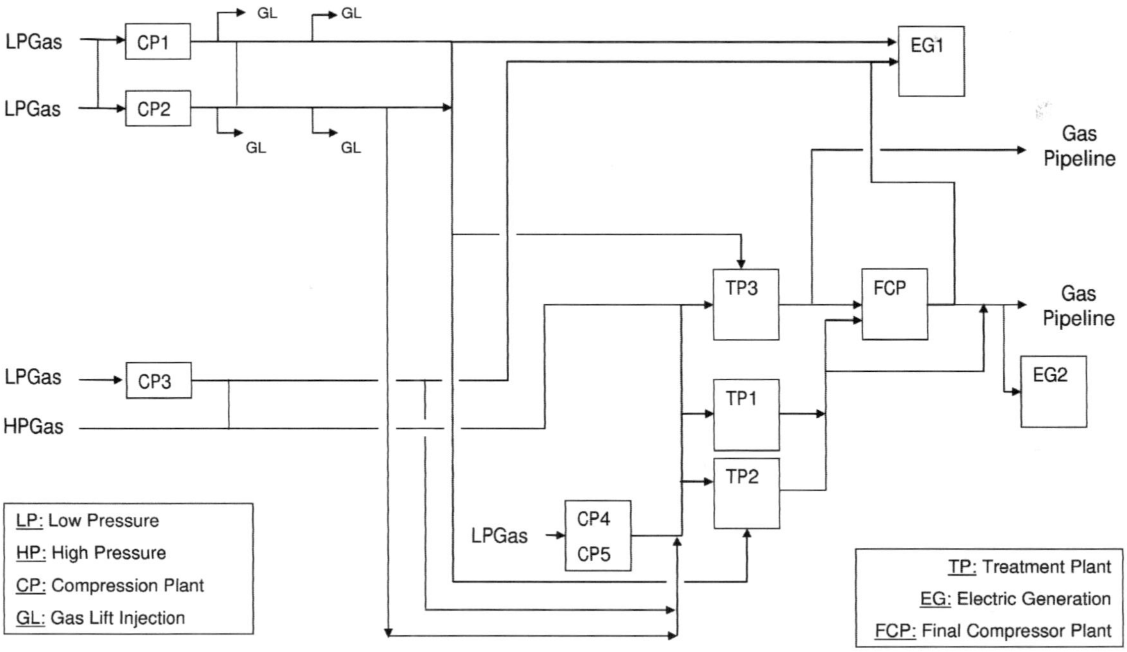

Figure 2. Detailed Gas System.

WORK PROCESS FLOWCHART

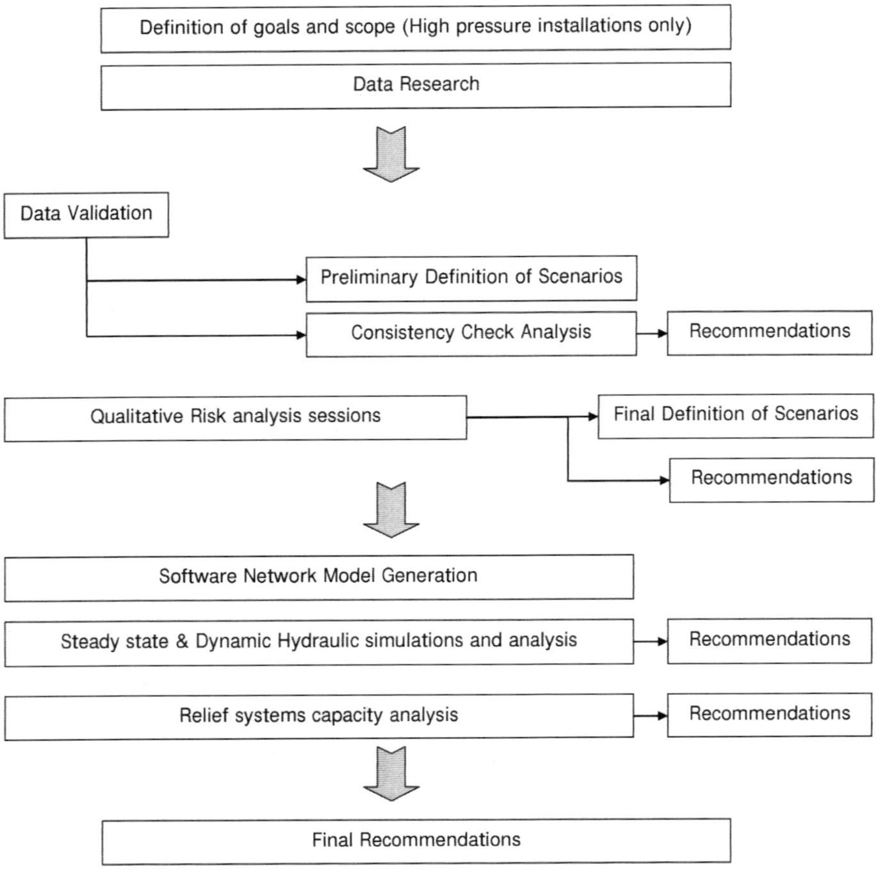

Figure 3. Work Process Flowchart.

DEFINITION OF SCOPE (PHYSICAL & CONCEPTUAL) & DATA VALIDATION

The Study focused on the high pressure installations, including five intermediate compression plants, one final compression plant, three gas treatment plants and more than 85 km of interconnecting gas pipeline, as shown in Figure 2. The main project goal was to analyze if existing protections of high pressure installations were suitable to guarantee a safe operation under a variety of possible scenarios.

Once the goal and scope of the project were defined, all data that was going to be used was validated. Validation of data was both complex and essential. All set points of alarms and safety elements were checked together with operational data and inconsistencies were resolved. Furthermore it was checked that previous (existing) HAZOP studies recommendations had been executed (or answered) and indicated on the proper related document.

During the project, two situations were studied, actual and future scenarios, taking into account probable changes in both flow and installations. In this presentation only the future scenario is shown, for simplicity reasons.

Main recommendations coming up during this stage: Adjustment of incorrect set points (safety element set pressures higher than design pressure) and errors in documentation.

DEFINITION OF SCENARIOS & QUALITATIVE RISK ANALYSIS

To analyze possible hazards and their consequences, a qualitative risk analysis study was organized.
A multidisciplinary team participated of in the study, including managers, operation representatives, process and safety engineers together with a safety consultant leader.
Main deviation of interest was high pressure. The system reviewed, as stated before, was the high pressure gas feed system to treatment plants 1, 2 & 3, electric generation & gas lift consumptions, and sales gas. Ten scenarios, with their consequences, safeguards and recommendations were analyzed. These scenarios were defined considering they were likely to occur in the lifetime of the facilities being studied.

- Scenario 1: Treatment Plant 1 Shutdown
- Scenario 2: Treatment Plant 2 Shutdown
- Scenario 3: Treatment Plant 3 Shutdown
- Scenario 4: Treatment Plant 1 & 2 simultaneous Shutdown
- Scenario 5: Treatment Plant 1, 2 & 3 simultaneous Shutdown
- Scenario 6: Final Compressor Plant Shutdown
- Scenario 7: Gas Pipeline 1 Blockage
- Scenario 8: Gas Lift Injection Blockage
- Scenario 9: Gas Lift Injection Blockage (Intermittent)
- Scenario 10: Compressor Plants 4 & 5 simultaneous Shutdown

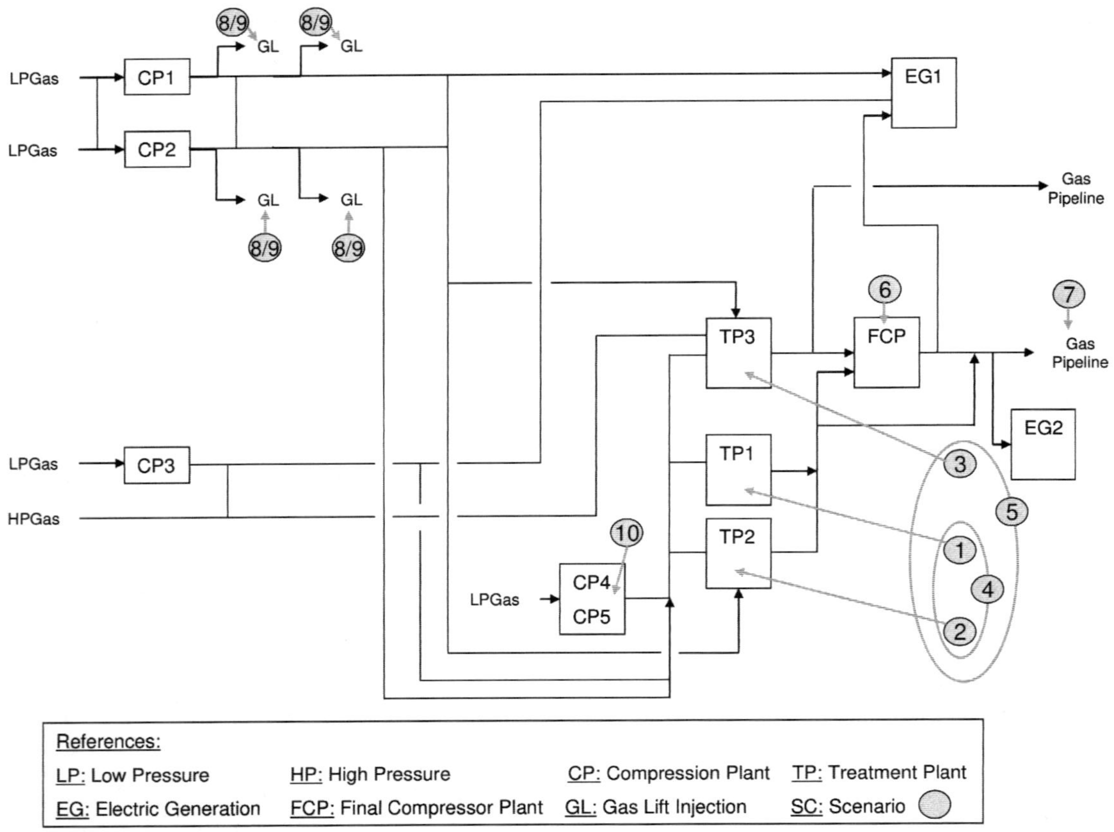

Figure 4. Definition of scenarios.

Main recommendations for this stage: Following a summary of results for scenario 4, as an example of the qualitative risk analysis sessions, is presented.

Table 1. Recommendations qualitative risk analysis.

Node: 2. Gas feed system to Treatment Plants 1, 2 & 3, electric generation & gas lift consumptions and sales gas. Future scenario.

Deviation: 1. High Pressure

Scenario	Consequences	Safeguards	Recommendations	Responsibility
4. Treatment Plant 1 & 2 simultaneous Shutdown	1. Compressor plants 4 & 5 Shutdown	1. Overpressure control valve in compressors discharge, Compressor Plant 1	10. Check new Facilities (installed downstream Compressor Plants 1 and 2, that will process the same compressed gas as the Treatment Plant 1) can manage possible rise in pressure due to Treatment Plants 1 and 2 simultaneous Shutdown	Process
		2. Safety valve (Full flow) in each compressor of Compressor Plants 1 & 2.		
	2. High flow in Treatment Plant 3 entrance (if entrance of Treatment Plants 1, 2 & 3 are connected).	3. Overpressure control valve in compressors discharge, Compressor Plant 2	11. Check flare system to be able to dispose simultaneously full flow and blowdown of Treatment Plants 1 and 2.	Process
	3. Possible Final Compressor Plant Shutdown.	4. Own compressor safety elements, Compressor Plants 1 and 2	12. Update of Cause and Effect Matrix.	Operations
	4. High Pressure in Compressor Plants 1 & 2.	5. Overpressure control valve in compressors discharge, Compressor Plants 4 and 5		
	6. Possible pressurization of Compressor Plant 3.	6. Overpressure control valve in compressors discharge, Compressor Plant 3		
	7. Rise in pressure and flow in new Facilities installed downstream Compressor Plants 1 and 2, that will process the same compressed gas as the Treatment Plant 1	7. Safety valve (Full flow) in each compressor of Compressor Plant 3		
		8. Compressor Plant 3 Shutdown		

ANALYSIS

In order to evaluate recommendations, three different studies were made:

- Steady state hydraulic calculations, with the intention of defining and validating the model to be used in dynamic calculations.
- Dynamic hydraulic calculations, so as to be able to evaluate system response to the different scenarios.
- Relief vent systems capacity analysis in order to assure no overpressure occurred in any of the installations.

These studies were made for all scenarios, except for scenarios 8 and 9. This was because during qualitative risk analysis study sessions, it was determined that these scenarios did not cause significant effects on installations, or any safety problems.
On the other hand, two new scenarios proposed by the team, were included: Compressor Plant 1 Shutdown and Compressor Plant 2 Shutdown.

In this paper only one scenario was chosen for simplicity. Scenario chosen is number 4 (Treatment Plant 1 & 2 simultaneous Shutdown).

Steady State & Dynamic Hydraulic Analysis

Steady state hydraulic calculations were made after loading the installations and pipe network into the software that was going to be used for both steady state and dynamic calculations.

In order to achieve this in a simple way but without losing the accuracy needed, certain simplifications were made:

- Compression Plants were considered constant flow contributions.
- Sales gas pipelines were considered as constant pressure nodes.
- Gas lift and electric generation were considered constant flow consumptions.
- Pressure drop in Treatment Plants were considered proportional to the circulated flow rate

Furthermore the following considerations were made:

- Constant and uniform Gas temperature for all network.
- Constant specific gravity of the gas.
- Pipeline roughness was considered 0.0009 inch.
- Shutdown valves were supposed to close at 1sec/in of diameter.
- Relief valves and safety valves were supposed to be fully open 0.5 seconds after reaching the corresponding set point.

With these considerations and simplifications, the steady state was run and results were validated with operational data and any deviations found were corrected.

Following, dynamic calculations were made to analyze how the system reacted to the scenario of Treatment Plants 1 & 2 simultaneous shutdown.

Table 2. Consequences of Scenario 4.

Event	Time	Consequence
1	8seg	High pressure alarm in CP4/ 5 discharge
2	8seg	Overpressure control valve opening in CP4/ 5 discharge
3	8seg	Safety valve opening in CP4/ 5 discharge
4	9seg	Overpressure control valve opening in TP3 due to high flow
5	9seg	Closure of shutdown valve in TP3 entrance due to high flow
6	51seg	Overpressure control valve opening in TP3 entrance
7	2min 50seg	Overpressure control valve opening in CP3 discharge
8	9min 54seg	High pressure alarm in CP3 discharge
9	15min 58seg	Safety valve opening in TP3 entrance
10	27min 27seg	Very High pressure alarm in CP3 discharge
11	27min 34seg	Closure of shutdown valve in CP3 discharge. Shutdown of CP3

Note 1: All times are referred to the initiating event instant. All simulations have its origin on the related steady state scenario.

Figure 5 shows the location and set points of pressure and flow rates of each safety element that was required to operate during the scenario 4 contingency, TP 1&2 simultaneous shutdown.

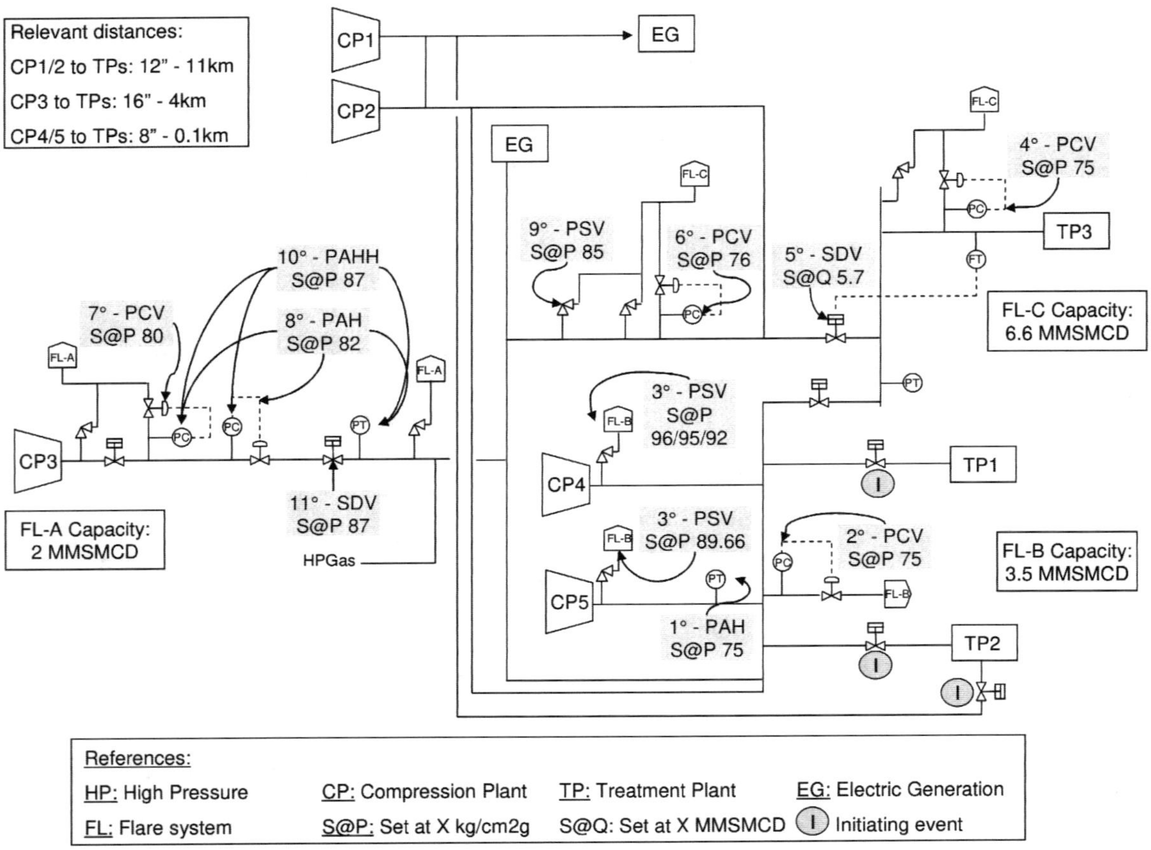

Figure 5. Consequences of Scenario 4.

Figures 6 and 7 show how the different safety elements open and how they discharge to the flare system.
Consequences detailed on Table 1 can be seen represented on the mentioned figures.

Taking into account that scenario 4 was evaluated in the most conservative way, considering TP1&2 shutdown, but with a continuous gas inlet of CP 4&5 (in spite of its own shutdown), TP3 shutdown is inevitably produced, due to inlet high flow rate.

As a conclusion from the dynamic analysis, it can be stated that in case of TP 1&2 shutdown, inlet gas must be restricted. Shutdown of compressors from CP 4&5 and closure of high pressure well has to be assured in order to avoid excessive gas flaring and TP3 high flow shutdown.

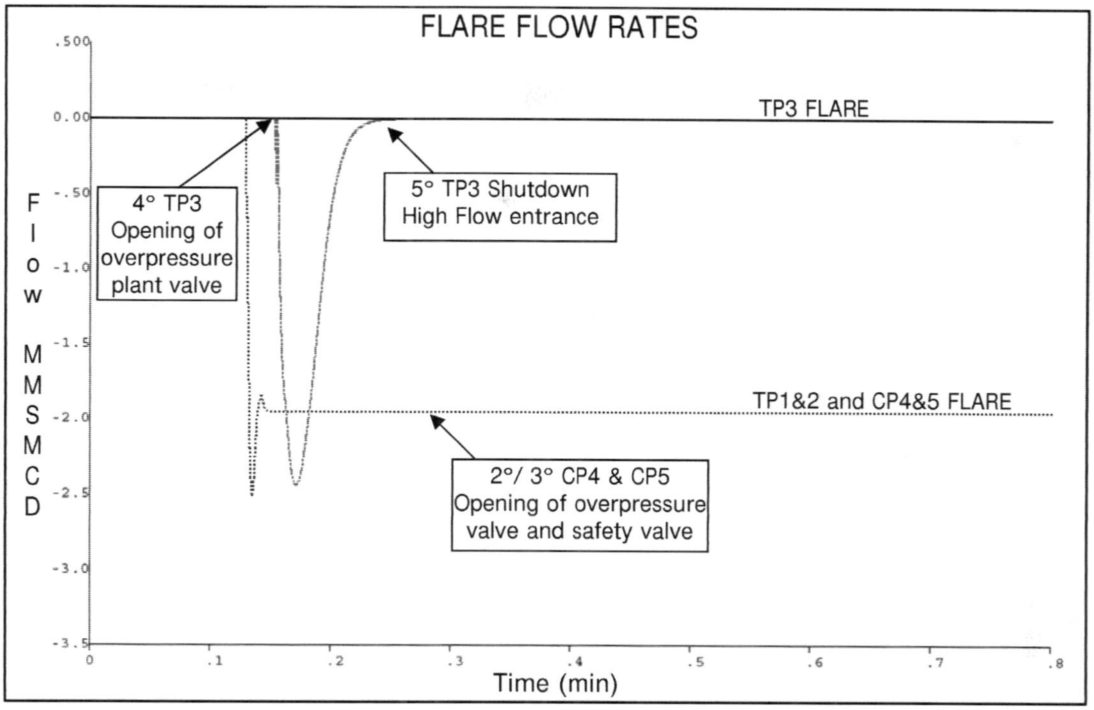

Figure 6[1]. Flare Flow Rates (Time 0-1 min).

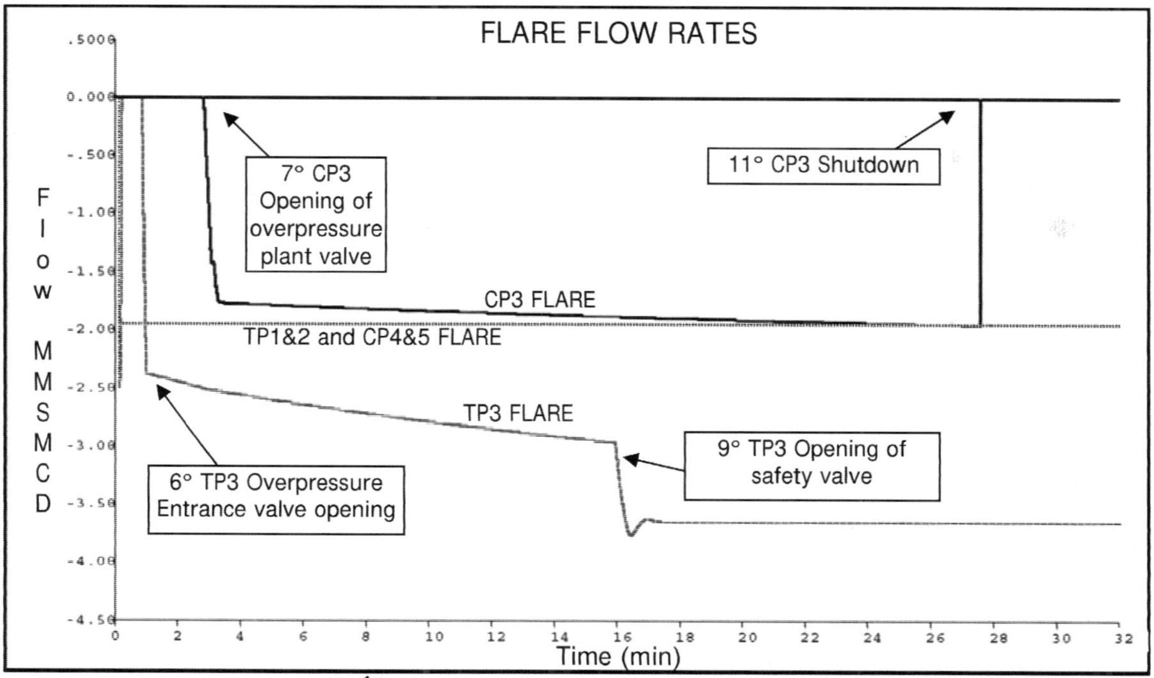

Figure 7[1]. Flare Flow Rates (Time 0-32 min).

[1] Figures 6 & 7 show negative flow rates since it's exiting the system (from the process to the flare)

Relief Systems Capacity Analysis

Due to changes in interconnections and increment of operating process capacities of Treatment Plants, considered in Future Scenario, the evaluation of the existent flare systems was required.

Two different systems were evaluated:

- Flare System FL-B: Treatment Plant 1&2 and Compressor Plant 4&5 (all facilities share the same flare system).
- Flare System FL-C: Treatment Plant 3

FLARE SYSTEM FL-B:

Different scenarios were evaluated:

- Gas inlet Full Flow (due to blocked outlet)
- Maximum flow of overpressure control valves and/ or safety valves (defined by the scenarios evaluated in the dynamic simulation)
- Instrument air system failure
- Emergency Shutdown of TP2 (with automatic blowdown). ESD of TP2 does not include other ESD of installations that share the flare system and continue to vent.

As a result of this study, one overpressure control valve failure position had to be changed from fail open to fail close, because in case of instrument air system failure the flare capacity was exceeded. This valve corresponds to the safety element that opens in second place on the Scenario 4 analyzed, see Table 2 and Figure 5.

Another conclusion of the analysis was that Emergency shutdown of TP2 with automatic blowdown could not be managed simultaneously by the flare system.
Blowdown must remain on manual mode or could be automated only if it can be guaranteed the feed gas flow rate from other installations is stopped. In this case, it was not an option to automatically delay depressurization, because system could not handle its flow.

FLARE SYSTEM FL-C:

Different scenarios were evaluated:

- Gas inlet Full Flow (due to blocked outlet)
- Maximum flow of overpressure control valves and/ or safety valves (defined by the scenarios evaluated in the dynamic simulation)
- Instrument air system failure

- Emergency Shutdown of TP3 (with automatic blowdown)

The main recommendation was the necessity of assuring immediate closure of high pressure wells, due to ESD of TP3 or instrument air system failure scenarios. This proposal was made to avoid overpressure and excess flow sent to the flare system.

FINAL RECOMMENDATIONS AND CONCLUSIONS

Different recommendations resulted from the multiple stages of analysis, mostly from consistency analysis and dynamic simulations.

From the consistency analysis, errors on alarm sets were detected and corrected allowing a better operating condition.
Furthermore, it was found some recommendations of previous HAZOPs were not fulfilled and an execution plan was established.
It is fundamental to keep documentation updated in order to assure all analyses made are reliable.

Main recommendations from dynamic analysis results were:

- Guarantee certain high pressure/ high flow wells shutdown when needed (system integrity response)
- Change in Set of PSV's

It is really important to analyze all installations together as a single system. This paper shows how a methodic analysis can contribute to understanding how our installations will behave under certain scenarios and how different events, occurring in one installation, affect others.
This analysis, allows evaluation of the integrity of our system, and teaches the people responsible for operation how their systems behave and how to react in case of an upset situation.

AUTHOR INDEX